BETWEEN SCIENCE AND TECHNOLOGY

North-Holland
Delta Series

NORTH-HOLLAND
AMSTERDAM • OXFORD • NEW YORK • TOKYO

Between Science and Technology

Proceedings of the International Conference
Between Science and Technology
Eindhoven University of Technology,
The Netherlands, 29–30 June 1989

Edited by

Andries Sarlemijn
Peter Kroes
Eindhoven University of Technology
The Netherlands

1990

NORTH-HOLLAND
AMSTERDAM • OXFORD • NEW YORK • TOKYO

ISBN: 0 444 88659 1

Published by:

North-Holland
Elsevier Science Publishers B.V.
P.O. Box 211
1000 AE Amsterdam
The Netherlands

Sole distributors for the U.S.A. and Canada:

Elsevier Science Publishing Company, Inc.
655 Avenue of the Americas
New York, N.Y. 10010
U.S.A.

Library of Congress Cataloging-in-Publication Data

International Conference Between Science and Technology (1989 : Eindhoven University of Technology)
Between science and technology : proceedings of the International Conference Between Science and Technology, Eindhoven University of Technology, The Netherlands, 29-30 June 1989 / edited by Andries Sarlemijn, Peter Kroes.
p. cm. -- (North-Holland delta series)
ISBN 0-444-88659-1 (U.S.)
1. Science--Congresses. 2. Technology--Congresses. 3. Casimir, H. B. G. (Hendrik Brugt Gerhard), 1909- --Congresses. I. Sarlemijn, Andries, 1936- . II. Kroes, Peter, 1950- . III. Technische Hogeschool Eindhoven. IV. Title. V. Series.
Q101.I623 1989
500--dc20 89-78346
CIP

Printed in The Netherlands

PREFACE

This book contains the papers presented at the conference *Between Science and Technology*, which took place at the Eindhoven University of Technology (The Netherlands) in June 1989. The conference was organised in honour of the 80th birthday of H.B.G. Casimir, the famous Dutch physicist who has made substantial contributions to the development of 20th century physics, and who was, for several years, head of an industrial research laboratory (Philips Research Laboratories). Outstanding scientists from different fields of research were invited to discuss important recent developments. The diversity of topics addressed in this book reflects Casimir's wide range of interests. The dominant theme, however, is the relation betweeen science and technology.

The editors would like to thank, in the first place, the section leaders of the programme committee of the conference: B. Bölger, L.-F. Feiner, J. Hilgevoord and H.A.M. Snelders; they greatly influenced the content of this book by their excellent selection of speakers. We would also like to thank the authors for their kind and generous cooperation in preparing these proceedings. Furthermore, we are very grateful to Louison Buijs and Pieter Gunter for their assistance in preparing the manuscript. Finally, our gratitude goes to Mark Eligh, Mary McAdam and Mary Carpenter of Elsevier Science Publishers B.V. for their advice and help.

Andries Sarlemijn
Peter Kroes

SPONSORS

The conference was sponsored by:

Philips International B.V.
Royal Netherlands Academy of Arts and Sciences (RNAAS)
Eindhoven University of Technology
University Fund Eindhoven
IBM
Océ – Van der Grinten N.V.
Foundation Physica
AKZO
NEDAP

TABLE OF CONTENTS

Part V: HISTORY OF INDUSTRIAL PHYSICS

Part I: Reflections on Science and Technology

Between Science and Technology
A. Sarlemijn and P. Kroes (Editors)

Chapter 1

Science and Technology – Present relations in historical perspective

A. Sarlemijn

University of Technology

Eindhoven

1.1 Siamese Twins

Relations between science and technology form the central theme of this book. These relations are strong, so strong that they are sometimes compared with those between married couples. Isn't it true that the relationship between a man and a woman develops only when they interact actively with each other? Isn't it true that, in a similar way, science and technology depend on the efforts and progress made by the other? Fundamental theories would remain fantasies if there were no possibility to test them with the most advanced modern technology as is done, for instance, at CERN, Berkeley, and DESY. On the other hand, stopping fundamental research is not compatible with a belief in technological progress. This strong interdependence suggests that the marriage metaphor might even be too weak. Statistics informs us about frequent divorces, whereas nobody would try to make modern science and technology independent of each other.

Perhaps a comparison with Siamese twins is more appropriate. A Siamese twin's liberty is restricted when s/he wants to go their own way. Don't we observe similar reactions in present-day research contexts? Managers of industrial corporations are often convinced that research workers should concentrate more on concrete products; they believe that these workers lack the desired insight into market evolutions. The impatience of employers is expressed through slogans like: "Inventions are not needed any more, innovations are required". From a short term economic point of view, the reason is clear: research costs vast amounts of money which should be paid back by market success, whereas research success always has a speculative character. From their more fundamental point of view, researchers are convinced that innovations depend on inventions, inventions depend on expertise, expertise depends on concentrated research, and concentrated research costs effort and time and, therefore, implies risks. And risk has never been excluded from the beginning of even the most important science-based innovations[1]. The impatient quest for innovations without risk leads, on the contrary, to frequent changes of focus and to the loss of an expertise which might be useful in future.

This book will not eliminate that tension between the twins. It presents instead an anthology of research. We hope it will stimulate reflections and lead to a better understanding of the richness of research, its prerequisites, its many sided relations with technology, and the innumerable directions in which research can be further developed. At the same time, the reader will get detailed information about important developments in special fields.

Chapter three in *the second part,* for instance, deals with a problem whose solution is desired in the context of many technologies: how to acquire theoretical insights into properties of materials. In addition, the chapter clarifies general characteristics of the transition from the necessarily idealized and formal research level to the domain of technological application. Martin F.H. Schuurmans reveals the fascinating results of industrial research in this field. He also makes plain the domain of the achieved insight: "The theory can only be applied to perfect crystals". In research contexts, idealizations make it possible to reduce the complexities of problems in order to find solutions for isolated aspects. Afterwards, reintegrating the problem in all its technological complexity, requires "good team work between computational specialists, physicists and chemists". Thus the chapter makes clear that theory application is not — as it is often believed — a mechanical activity which can be done by "engineering without meeting fundamental problems". The chapter not only gives a thorough insight into

[1] For Casimir in [6], this aspect of innovations forms the paradox of research management: "we should aim at success, but if all our plans are successful, we have been poor research managers".

an important evolution in a special field, but also illustrates a characteristic feature of the application of research results in technological contexts.

Joseph E. Demuth's contribution too contains detailed information together with more general conclusions. It presents, in a clear and illustrative way, the scanning tunneling microscope and the interesting things which we can observe with it. But, title, introductory paragraphs and further context clarify the author's more general aim: "revolutions" in modern scientific disciplines do not belong — as prejudice would have us believe — to the exclusive privilege of universities; they sometimes originate from industrial research as is the case with this new tool developed by IBM scientists; it allows one "to directly probe surface geometric and electronic structure on an atom by atom basis", whereas, in the past, macroscopic measurements were used to "infer" the characteristics of the surface on the atomic level. This implies a completely new look at things and also a new way of working in a scientific discipline.

Following the chapters on industrial research is *the third part,* which is on fundamental physics. One might ask how this part fits into the general theme of the book. Indeed, "pure scientists" sometimes neglect technological relevance or mention it as an "afterthought". Discussing the final aim of theoretical physics at a special occasion[2], Gerardus 't Hooft acknowledges the speculative character of this subject: none of us will be around at the realisation of that final aim. Nevertheless, in his opinion, it could be said that theoretical physicists want to understand "how physical laws are connected with each other and what we could do with them". Given this fact, the final aim could, in a certain sense, be defined as the so called "Theory of Everything": that is to say "a mathematical theory about all natural phenomena" or an equation which makes it possible to calculate — like Laplace's superintelligence — everything under all circumstances. The solution of this equation will give rise to problems. But at least it will allow us to arrive at approximations. Indeed, within this consideration, solving technological problems is not of prime concern.

Sophisticated reflections, however, cannot allow us to bypass dealing with "pure" physics; the latter can be considered as a contrast to technology and as a necessary point of reference. Only if its autonomous function is recognized, can its relations become a fascinating problem, because even in its "purest" form, science is not completely isolated from technology. It has already been mentioned that the testing of modern physical theories depends on advanced technologies. We come across a special kind of this dependence in John S. Bell's chapter dealing

[2]G. 't Hooft interviewed on 11.7.1989 by W. Kayzer (available on cassette from the "publiek-service" of the VPRO, Hilversum).

with the Einstein–Podolsky–Rosen paradox: several thought experiments of half a century ago have now become into experiments which can really be done in specialized laboratories.

The papers in *the fourth part* unlock new fields of research. One of them is *synergetics* which analyses in an interdisciplinary way the combined behaviour of different particular systems. The combination of these particular systems cannot be described satisfactorily with the entropy concept and conservation laws, because it does not form a closed system; therefore, new concepts have to be developed. The first to work on this task has been Hermann P.J. Haken. In encyclopedias, for example in [1], his name is immediately associated with this new field. His contribution shows how synergetical methods make it possible to predict quantum statistical properties of laser devices. With examples and illustrations, he also indicates how *synergetics* can lead to the development of machines for pattern recognition. Whereas Haken develops ideas for new technologies starting from a more fundamental approach, Manfred Eigen borrows a concept from physics (the concept of phase changes) and applies it to information transport to arrive at fundamental insights into the evolution of life. He shows that new physical concepts and laws are required, if we want to deal in an exact way with the generation and the transformation of information in the evolutionary process.

The fifth and final part is dedicated to history. It has often been said that historical studies do not lead to a valid pattern for *all* innovations or inventions. Generally speaking, this objection holds because there are too many differences between the various technologies. At first sight, the reader might get the impression that this objection can also be levied at Lillian Hoddeson's aim: her contribution presents a pattern of three phases of inventions (and compares it with Casimir's model[3]). But, we should not forget that she concentrates on *the special kind* of research which has led to the inventions of the repeater and transistor in the Bell Laboratories. These inventions constitute prototypes of science-based innovations. Such innovations could only be achieved through tentative designs and constructions carried out in close collaboration with scientists who were familiar with fundamental problems, like Bardeen who, taking off from from Shockley's unsuccessful design of a field effect transistor, discovered the surface states of electrons in semiconductors (Demuth too mentions this example). This collaboration forms the second phase of Hoddeson's pattern and it is this phase to which she pays the most attention. It goes without saying that the "pattern" of collaboration would have

[3]Casimir's analysis of transistor research can be found in [3], especially on pp. 87–88. He uses it as an example to show that "industry depends on the results of fundamental research — and here [he means] to say Fundamental research with capitals — but that such results do not come out of industrial laboratories".

been completely different, if a special design of a corkscrew or a special house had resulted from the "research" and not a microtechnological product. Therefore, the last chapter can be seen as a detailed description of the necessary coordination of specialisms which are required to achieve microtechnological innovations like that of the repeater and transistor.

We may conclude that various types of relations between science and technology are discussed in this volume: the complexity of the application of a special theory in technology (Schuurmans); the impetus to fundamental new ways of working in a scientific field born within an industrial context (Demuth); the richness of "pure" science ('t Hooft, Bell); the new technological ideas based on a new fundamental approach (Haken); the possibility of arriving at new fundamental insights starting from modern technologies (Eigen); and the need for collaboration between "pure" scientists and "technologists" in order to achieve microtechnological innovations (Hoddeson).

1.2 A retrospect

We are inclined to think that our urgent problems are unique. On critical consideration, this often turns out to be untrue. That is also the case with the present tension between science and technology: it is not unique.

(a)

Centuries before Christ, the Alexandrian school refused to follow the fundamental approaches of the other Greeks and wanted to concentrate on practical and exact results. The Alexandrians used a handsome distinction to apply mathematics in fields like mechanics, hydrostatics, optics, and astronomy without dealing with fundamental questions. They considered these disciplines as not being scientific activities but arts; at the same time, they encouraged the concentration of academic activity on "arts" instead of on "science". The trick led to impressive results. Even today, children still learn the insights of Archimedes, who was educated in Alexandria. Useful engines were designed in the town and even steam found its first technological application there. Only alchemy was developed in a speculative way. From our present knowledge, at least one of the reasons for this speculative development is clear: the scientific and technological presuppositions to realize the aim of alchemy — making gold or silver — were far from achievable in that time. However, the example of the Alexandrian school illustrates how old the tension between fundamental approach and the desire for exact and practical results is.

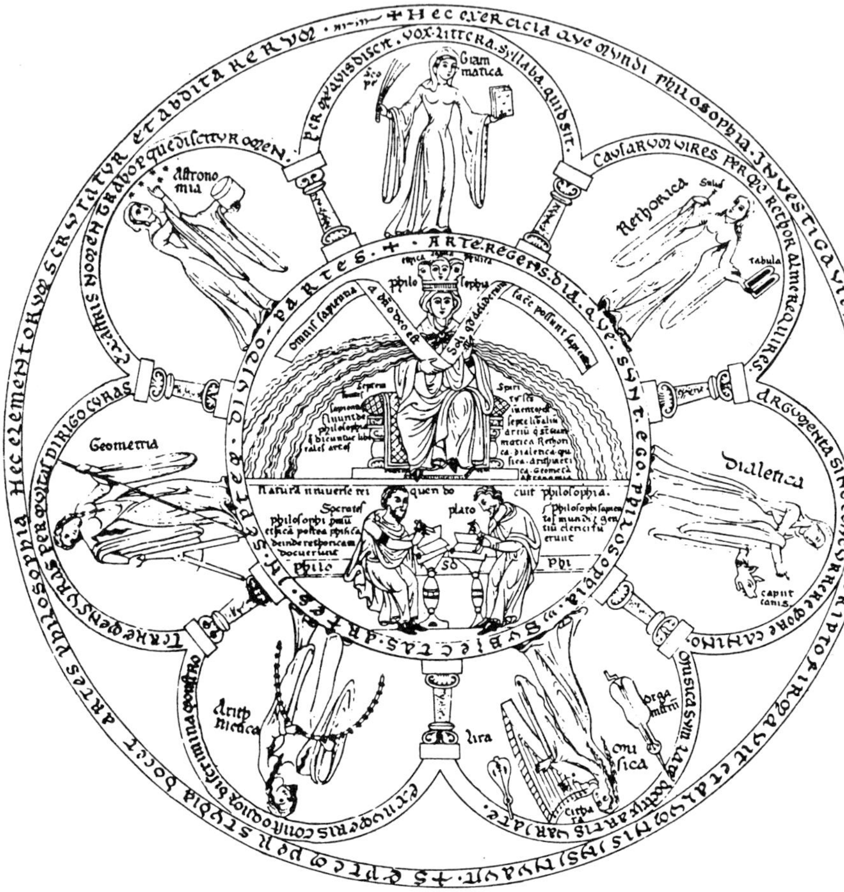

Figure 1.1: *Part of a drawing (from 1190) showing the relation between sciences and arts. Above the crown of the sitting person three sciences are mentioned (physics, logic and ethics); the sitting person radiates wisdom onto the seven arts whose names are written between the arcs: the mathematical arts (astronomy, geometry, arithmetic and theory of harmony) and the linguistic arts (grammar, rhetoric and dialectic).*

In the Middle Ages, similar distinctions were maintained: mechanics, optics and astronomy were considered arts while alchemistic and medical problems belonged to the practical and nonacademic disciplines. Only Albertus Magnus and Roger Bacon recognized a scientific alchemy and set it against practical activities. To illustrate the usefulness of scientific alchemy, Bacon introduced a nice allegorical comparison which can perhaps still be used in discussions. He thought that the work on scientific alchemy could be compared with the efforts of sons whose father had told them before his death that a precious treasure was hidden in his orchard. The sons could not find the treasure. But, in searching for it, they ploughed the orchard thoroughly. The result was that in the following spring, the trees blossomed abundantly. Then they realised that they did not possess a treasure in a strict sense but a rich source of wealth. The history of science contains many such comparisons about the relations between science and practical results.

(b)

Newtonian mechanics, thermodynamics and the theory of electromagnetism are the main successes of the classical period. Our present technology is still largely based upon macroscopic physics, the basic equations of which were found in that classical period[4].

How could these impressive results be achieved, when the preceding period did not bring anything comparable? Our textbooks tell us that the only new things Kepler and Galileo did were observing and measuring phenomena. I think that we should no longer permit such fallacies about the combination of science and technology to be told in school. In fact, Kepler, Galileo and their contemporaries did something completely different.

Although strongly linked with each other, *two aspects of the new approach can be distinguished:* (b.1) the classics eliminated the ancient demarcation between science and arts or technological disciplines and (b.2) they ignored the problems discovered by the Greeks in the context of the foundation and application of mathematics.

(b.1)

Copernicus, Tycho Brahe, and Kepler presented their astronomical insights as a kind of mechanics, and this mechanics has since then been considered as belonging to physics. The new approach is even clearer in Galileo's *Discorsi* (1638). Galileo presented mechanics and physical foundations of ballistics as "new" sciences, whereas at that time Simon Stevin was still considering his mechanics in the traditional way — as an *art* of weights ("Weeghconst"). The new

[4]See chapter IX of [4] and [5].

aspect was that Galileo no longer considered these disciplines as arts but as physical sciences concerned with physical laws e.g. on pulley-blocks and on falling bodies and projectiles.

Kepler, Galileo and their contemporaries developed insights which later became the "material" of the "synthesis" of Newton. It was Newton who introduced the method of axiomatic deductive explanation in a field of physics to which not only "untouched" physical phenomena but also "artificial" motions of engines belong. *This was a new method in physics. It became characteristic for the classical period.*

The birth of thermodynamics occurred in a similar way; Carnot's *"Reflexions on the Motive Power of Fire and on Engines Fitted to Develop that Power"* (1824) begins with the "proof" that the "motive power" of the "heat engine" is a natural force:

> "It is generally known that heat can be the cause of motion and that it possesses great motive power. The steam engines in widespread use today are visible proof of this. We must attribute to heat the great movements that we observe all about us on the Earth" ([2], p. 61).

With these sentences began the search for basic concepts and equations for natural and artificial heat phenomena. Maxwell's basic concepts and equations for electromagnetism did not need such a "proof" after Ørsted's discovery of the unity of magnetism and electricity; magnetism has been considered a natural phenomenon ever since Gilbert's book on this topic in 1600; and electricity has always been considered as an artificial and natural phenomenon since Guericke's experiments.

Therefore, we can say that the three main classical theories are based on the application of the axiomatic deductive method in order to explain not only natural but also "artificial" phenomena.

(b.2)

Let us now pass to the second aspect. At the beginning of the classical period, the problems discovered by the Greeks in the context of the foundation and application of mathematics were neglected; therefore, little attention was paid to the real reasons for the ancient separation of science from arts or technological disciplines. In this sense, the two aspects (separation of science from technology and fundamental mathematical problems) are interconnected.

Why did the classics relinquish the ancient distinction between science and arts? The arguments were not based on the experimental method, the observing and measuring of phenomena or the application of mathematics to find exact formulations of laws. These approaches were not new (only Francis Bacon, who knew little about mathematics and other arts, thought that they were new). They had been characteristics of mechanics, optics and astronomy and had formed the very reasons why the ancients had not included these disciplines among the sciences.

Ancient mechanics was experimental and exact, but it did not claim for itself the ability to explain natural phenomena. Handsome tricks formed its aim as it still does in the art of accountancy. A μηχἄνή was a handsome instrument like a pulley; the ancients thought that with it only an artificial motion could be realized. Galileo objected that while mechanical instruments could be handsome constructions, they were, nevertheless, based on natural motions and that they were restricted by natural impediments and limits. Therefore, he postulated that motions of mechanical instruments too should form the subject of physical explanations. He formulated a similar postulate for the discipline on falling bodies and projectiles. In this sense, he presented his two "new" sciences on mechanics and on physical foundations of ballistics.

But, this is still only a part of the history. In order to understand the whole story, we have to deal with the question of why the tasks of the ancient disciplines like mechanics were limited by the Greeks. They had good reasons for it. We must not forget that they were brilliant mathematicians; their approaches in physics and arts were influenced by mathematical insights.

They discovered that mathematization presupposes idealization and that idealization leads to properties of the formal subjects — like geometric incommensurability, infinity and discontinuity — which cannot be ascribed to their real counterparts. This discovery had consequences for the ways of dealing with mathematical subjects itself: infinities were avoided as much as possible; proportions and irrational numbers became suspect; proofs of equations and those of geometrical theorems followed the complicated methods of geometrical algebra.

With regard to the application of mathematics in physics, the discovery led to three different approaches: the *atomists* ascribed incommensurability, infinity and unbounded divisibility only to mathematical subjects and denied that physical subjects would possess similar properties; *platonism* refused such a concession and emphasized that scientific knowledge should imply a preference in favour of idealized forms; but, more influential was the approach of the *aristotelians* who, against platonism, quoted the properties mentioned before as arguments to show that essential forms of real motions cannot be described by mathematical means. Therefore, aristotelian physics dealt with the "real" essences of motions, considering mathematical idealizations as a "lower" form of abstraction. For the same reason, physical explanations were reserved for "science" and the "naive" application of mathematics was left to (practical) arts.

Galileo's arguments to give up this demarcation have already been mentioned. But neither he, nor Descartes, nor their contemporaries solved the mathematical problems discovered by the Greeks. Hans Freudenthal has often emphasized — for instance in [7] and [8] — that the classics used "free of care and merrily" mathematical infinities although the methods to deal with them were not developed prior to the 19th century. Therefore, an interesting question is how the classics dealt with the gap between idealized subjects and their real counterparts. In my opinion, *there are four different reactions:* that of Galileo and Newton; and that of rationalists and empiricists.

First reaction: Against aristotelism, *Galileo* postulated that essential forms of motions can be formulated by mathematical means. Nevertheless, he did not deny the opposition between idealized and real forms[5]; this gap could be bridged, he thought, by distinguishing three tasks: the finding of right idealizations is the task of *mathematical physics* in cooperation with *experimental physics*, which has to find experimental situations in which the idealized phenomenon is strongly approximated; "disturbing" factors are left to *technological arts*, because these factors form an "infinity" of unfixed variations and scientific research cannot be done on unordered infinities. In this sense, the distinction between science and technological art was still supported by Galileo. The only important change was that mathematical mechanics now belonged to physical science and was no longer considered an art.

Second reaction: *Descartes's rationalism* has had two sides. He wanted, first of all, a causal explanation of the laws of the falling bodies and he wanted to know if these laws were valid. To find the explanation he introduced his famous theoretical model of vortices of particles. In his opinion, this model made it possible to conceive, for instance, the heavenly motions as mechanical ones, like those of a machine; the clock became the cartesian prototype of a machine. Secondly, Descartes wanted an axiomatically deductive treatment of those laws from more general principles. To arrive at this deduction he introduced the inertial principle. These two sides of cartesian rationalism made it possible to bridge the distance between reality and idealized subjects in the context of the deductions of equations: the theoretical model had to lead to a clear and distinct representation of those idealized subjects.

[5]Therefore, Galileo must not be considered as a platonist (like some historians do).

Third reaction: *British empiricism* refuses to assume theoretical entities like cartesian vortices; it also requires prudence with regard to idealizations presupposed by mathematizations. Perhaps, Boyle's approach can be quoted as the one which approximates very closely the ideals of empiricism. If we look at the way he presented his law, then we see how he reduces the influence of theorization and idealization to the minimum: he provides us with a detailed description of the data on which his equation is based. In his presentation, the mathematical equation has no further status than being an *"abbreviation"* of many empirical facts; connections with theoretical assumptions are denied.

At the end of the last century, Ernst Mach tried again to reduce the influence of idealizations and theorizations. Physicists, like Albert Einstein and Felix Ehrenhaft, were impressed by his fervour in trying to convince us that we have to be cautious and prudent in the context of the empirical foundation of physical laws. An argument in favour of this empiricism is the fact that equations like that of Boyle should — if possible — be substituted by better ones and that our dogmatic belief in discovered laws should not form a barrier against new insights; indeed, Van der Waals's equation corrects that of Boyle regarding those facts which Boyle could not have observed because of his limited technological means. Mach's fervour is impressive, but we should not overlook the difference between the context of empirical foundation and that of mathematical physics. Exceptions or uncertainties must be accounted for in empirical contexts. On the contrary, in mathematics, equations are either proved or unproved; "half-proofs" and "exceptions" do not make sense; they do not occur within a mathematical approach. Therefore, mathematical physics would become impossible, if empiricism were always followed in a very strict way.

Fourth reaction: The cartesian ideal of mathematical physics based on the axiomatic deductive method was realized by *Newton* in his *Principia* of 1687. With regard to idealizations, he followed neither Galileo nor Descartes. The "preface" of the *Principia* was a violent reaction against the galilean opinion that there are two kinds of mechanics: a mathematical and a technological one. It would be absurd to think, he emphasized, that practice would lead to laws which differ from the mathematical mechanical ones.

> "The ancients considered mechanics in a twofold respect; as rational, which proceeds accurately by demonstration, and practical. To practical mechanics all the manual arts belong, from which mechanics took its name. But as artificers do not work with perfect accuracy, it comes to pass that mechanics is so distinguished from geometry that what is perfectly accurate is called geometrical; what is less

so, is called mechanical. However,the errors are not in the art, but in the artificers" ([10], p. XVII).

The second book of the *Principia* deals with the task which had been reserved by Galileo for practical mechanics; the infinity of "disturbing" factors becomes reduced to only one kind of disturbance: resistance. The book analyses different phenomena and demonstrates the validity empirical laws describing resistance.

With regard to Descartes's vortices, Newton is more in agreement with British empiricism than Boyle, who still considers them as a handsome hypothesis. At the beginning of his career, Newton too had accepted this hypothesis. But, some years before the publication of the *Principia* — scholars disagree on the exact date — Newton carried out his pendulum experiments and came to the conclusion that no mechanical influence could be ascribed to the cartesian particles. Newton's "hypotheses non fingo" can be interpreted in many different ways; one of them is that "pure" mathematical physics was made possible by Newton's rejection of the cartesian hypothesis. Indeed, this interpretation has been defended by d'Alembert and Laplace.

That it is difficult — as we have seen — to combine mathematical physics with a strict empiricism becomes also clear from the fact that nobody took seriously Berkeley's objection to Newton's concepts of time and space. From the point of view of empiricism, Berkeley was right. These concepts do lack an empirical foundation. Mach was the first who quoted the objection at the end of the 19th century. Mach's merits for experimental physics — clarified by expressions like "Mach number" — are not a subject of any doubt, but mathematical physics was not his primary interest.

Summarizing, we can say that there are at least *four "philosophies" on physical idealization*. The danger of false idealizations can be reduced by the approach of *empiricism* namely by keeping in mind the extent to which some empirical facts are at variance with the equation and that some phenomena have not been observed. In the context of mathematical physics, it is often easier to follow the *galilean* approach, keeping in mind one prototype of an experimental situation which approximates the idealized situation. If we follow the *cartesian* method, then we are looking for a theoretical model as a visualizable but not strictly empirical representation of a more fundamental process (like that of the vortices) to explain the phenomena under consideration. The *newtonian* approach in mechanics eliminates theoretical concepts if they lack an experimental content, and it reduces the distance between idealization and real facts by generalizing the "disturbing" factors and by finding laws governing those factors. But, none of these approaches eliminates the gap or leads to a definitive answer to the question of how to deal with the gap in all situations.

(c)

These "philosophies" are not the result of "pure" thinking. They correspond with attitudes. Physicists determine their attitude depending on the stage reached by research into a problem. Therefore, it is possible that a physicist follows all these "philosophies" during his career.

The history of quantum theory delivers a nice example. In the twenties, Bohr, Sommerfeld and Kramers worked on the well-known theoretical model representing the atom as a "planetarian" system. Defending a kind of empiricism, Heisenberg and Pauli attacked the model and viewed "the assumption of the existence of electron orbits" as a "prejudice" without empirical foundation. After one of Bohr's visits for a lecture in Göttingen, for a so called "Bohr festival", Heisenberg remembered how Einstein's criticism on classical mechanics had been influenced by Mach's empiricism. Therefore, Heisenberg expected approval of his empirical rejection of the quantum theoretical model. But the attempts to turn Einstein into his ally were in vain. The Einstein who started to work on the theory of relativity no longer existed: from a "philosophical" point of view he had changed his mind because he had discovered that, in the context of quantum theory, the discussions on the atom model could not be based on strictly empirical concepts.

However, Heisenberg had not yet given up his empiricism. He continued to work on his famous article *"Quantum-Theoretical Re-Interpretation of Kinematic and Mechanical Relations"* and finished it just before he left for a lecture at Cambridge in 1925. He himself did not know what to think about the content of the article. Therefore, he entrusted Born with the text. After having read it several times, Born discovered that its content could be presented in an exact deductive way with the matrix calculus. He entrusted his assistant Jordan with the task of working out the matrix mechanics. But, at first, Pauli and Heisenberg reacted in the same negative way to the characteristic idealizations of Born's matrix mechanics as they had done earlier to Bohr's theoretical model. Pauli even reproached Born:

> "You are only going to spoil Heisenberg's physical ideas by your futile mathematics".

But, a year later, all objections were dropped and Heisenberg himself began searching for an appropriate theoretical model for the matrix mechanics[6].

[6] A more detailed analysis can be found in [9] and [12].

These events inform us how physicists choose or give up their "philosophies" depending on the stage of the problem and also depending on the aspect of the problem which they want to study. Einstein did not accept Heisenberg's point of view, although he had held a similar standpoint earlier; we can understand his attitude, because at that time he was hardly interested in the empirical problem of quantum theory which bothered Heisenberg. "Positivism" led to the insights of Heisenberg's article. He and Pauli could not accept the theorization of Bohr's model and the idealization of Born's matrix mechanics prior to 1925, because then they still doubted the crucial facts. But, when Born and Jordan had accomplished their formal work, Heisenberg and Pauli saw how their empirical questions could be solved with the calculus. Then, the doubts disappeared and the way for theorization became free.

The previous analysis shows one of the many difficulties concerning the manipulation of research and of the growth of knowledge to make it more efficient. Doubts about the sense of theoretical models or of idealizations can be well founded. This is especially the case if facts are problematic.

In the context of technology too, it is sometimes extremely difficult to judge idealizations in the right way. The Teal story shows that even a genius like Shockley could make "mistakes" on this point. In 1948, as a member of the Bell Laboratories, Gordon Teal invented the "double doping" or grown junction single-crystal technique, which made possible the mass production of transistors. He joined Texas Instruments in 1952 and built up the company's research laboratories, making possible the production of the first commercial silicon transistor. He had a "personal sentimental attachment" for germanium long before anybody had thought of a transistor. Both his Master's thesis in 1928 and his doctoral thesis in 1931 dealt with this material which had been fascinating him throughout his graduate school days:

> "It was a material studied only for its scientific interest; its complete uselessness fascinated and challenged me" ([19], p. 621).

In the forties, it had still been difficult to know in advance that Teal's theoretical insights would lead to success. As soon as the point contact transistor was invented in December 1947, Teal went to Shockley to explain his thoughts about a new process to produce germanium. He was convinced that polycrystalline germanium with its variations in resistivity and its randomly occurring grain boundaries, twins, and lattice defects that acted as uncontrolled resistances, electron or hole emitters, and traps would affect the transistor operation. Therefore, Teal thought that the high purity of an ideal crystal should be approximated as much as possible

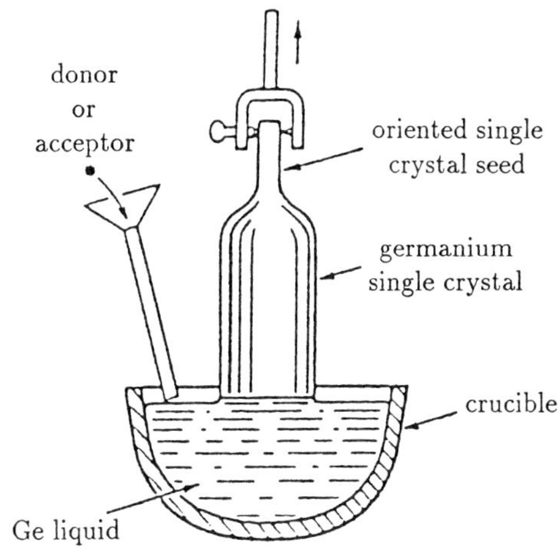

Figure 1.2: *Schematic drawing of Teal's technique of making p–n junctions and junction transistor structures [19].*

during the production. But, Shockley wanted to save time and attention for other things like his idea on field effect transistors. He was also opposed to the work on single crystals because he thought that transistor science could be elicited from small specimens cut from polycrystalline masses of material. Teal did not get permission to work on his ideas and had to continue with his earlier task: the study of chemical aspects of silicon carbide varistors for a new telephone handset. In secret, he designed the equipment for crystal growing; and it started functioning in 1948 (see fig. 1.2). A year later, he was able to make p–n junctions by doping with donors and acceptors during the growth of the crystal. At the beginning, Teal's disobedience had been — as Shockley's report [16] of 1974 tells us — risky, because he could have lost his job. But as early as 1951, Shockley recognized that he had been wrong and published, together with Teal, the article [18] on germanium p–n junctions.

The bulk production of transistors, however, was preceded by a discussion about the technological relevancy of an idealization. After the explanation of his equipment and the production processes invented by him, Teal concludes:

> "Actually, it should be emphasized that these physical pictures with which we work and upon which theories must be built are highly idealized and assume a perfection in structure and control of purity and composition of germanium and silicon that were not present in the actual materials under study until the work discussed here was done" ([19], p. 627).

From all this it becomes clear that *even the most technologically relevant research depends on a decision about "philosophies" on idealization.*

These decisions seem to form an essential part of the progress of "pure" and applied science. They also make it difficult to turn scientific research into an assembly-line activity.

1.3 Back to the present

Since the middle of our century and, therefore, since the time transistor research began, many things have changed. The change not only had scientific, but also social dimensions. If we want to deal with these aspects, we can — as Casimir does in [3] and in footnote 3 — distinguish between Fundamental research with capital "F" and fundamental research with a small "f". After the setting up of the industrial laboratories at the beginning of our century, the fate of the last-named research was linked to the evolution of industrial concerns.

Let us first consider research with the capital "F". Einstein's general theory of relativity and his later theory about the unity of the basic physical forces belong to this research. Formulating these theories demanded varied abilities like intuition, mathematical knowledge, insight into relevant experiments, and so on. The same abilities are still required in the context of high-energy physics. But something more is needed now. There must be a link with an advanced and extremely expensive technology to test the theories. The enormous amount of money involved is the reason why democratically elected governments require justifications for this kind of research; at the same time, idealizations and theorizations have a form which is extremely abstract and far removed from our daily life; therefore, it is difficult to give the information required by governments. This is a new phenomenon in history.

The situation with regard to the second kind of research has changed too. By the twenties AT&T, General Electric, DuPont, Eastman Kodak in America and AEG, Siemens and Philips in Europe were willing to pay for their own fundamental research. The employers and managers of the concerns still recognize the relevance of this research. But, they suspect that research workers do not sufficiently realize the Siamese connection: living apart together is not allowed. The economic situation, moreover, increases the pressure on the research to demonstrate its usefulness. But this is not easy because various circumstances have changed. I will mention some of them.

First: Research dealing with the electric lamp, the telegraph, the telephone and so on was, in a certain sense, speculative in that time. But their financing was no problem. Nowadays, such projects require — not only absolutely but also relatively in the context of annual turnover of industrial concerns — much more investment, as the Philips Mega-project shows.

Second: In the twenties, the possibility of bulk production was envisaged. Nowadays, it is difficult to find a link between research and a clear gap in the market for a new kind of mass production.

Third: During the first half of our century, basic physical research, especially the discovery of the microphysical world, was clearly directly relevant for the products of that time, and many results confirmed this expectation. Nowadays, nobody will deny that high energy physics has fewer connections with practical questions than Thomson's research on electrons and Kamerlingh Onnes's low temperature physics.

Fourth: Casimir has introduced the distinction between "F" and "f". In his time, the distinction was more or less clear: at the end of the twenties, quantum mechanics had found its final mathematical form and in that time the practical relevance of parts of this theory was intuitively clear. It has become difficult for a company, nowadays, to judge the many possibilities to do research and to select the possibility which has a practical relevance for its production. Why should it take all the risks? Managers are more and more convinced of the idea that every fundamental research (with minuscule "f") of interest for many industries should be financed by the state or (in Europe) by the European Community.

Many other circumstances have changed. The main change comes from the hard competition on the market. The struggle to survive is much stronger now than was the desire to expand in the twenties. The Japanese challenge makes people nervous; reflections and prudent judgements are not easy in such a situation. So it seems to be an appropriate moment for the academic world, and especially for the universities, to pay attention to these questions in more detail than was done earlier. But in some countries, the nervousness in industry has infected the academic spirit[7]. From that point of view, this volume represents a reaction against the common trend; it is based on the belief and on the knowledge, which have inspired Casimir throughout his life: he believes in the fact that research creates new concepts which change our world view and our way of living, he knows that successful research (as our trip through the history of physics has shown) depends on decisions about "philosophies".

[7]Much more can be said about the new situation. But a detailed research management analysis does not fit into the general character of this book.

Acknowledgement

I am grateful to Dhun Prasad, Peter Kroes and Louison Buijs for their comments on a draft version of this article and for their help with the English language.

References

[1] *Brockhaus; Natuurwissenschaften und technik*, Bd. IV, Brockhaus, Wiesbaden, 1983.

[2] S. Carnot, *Reflexions on the Motive Power of Fire and on Engines Fitted to Develop that Power*, UP, Manchester, 1986.

[3] H.B.G. Casimir, Philips Technical Review **20** (1958/59) 85.

[4] H.B.G. Casimir, *Haphazard Reality*, Harper and Row, New York, 1983.

[5] H.B.G. Casimir, *De kringloop van de natuurkunde en techniek in de 20e eeuw*, Hollandse Mij. der Wetensch., Haarlem, 1979.

[6] H.B.G. Casimir, Physics Bulletin **33** (1982) 314.

[7] H. Freudenthal, "Pythagoras", in: *Die Grossen der Weltgeschichte*, Teil I, Kindler Verlag AG, Zürich, 1971, pp. 332–353.

[8] H. Freudenthal, "Euklid", in: *Die Grossen der Weltgeschichte*, Teil I, Kindler Verlag AG, Zürich, 1971, pp. 710–725.

[9] J. Mehra and H. Rechenberg, *The Historical Development of Quantum Theory*, Vol. 2: *The Discovery of Quantum Mechanics 1925*, Springer, Heidelberg, 1982.

[10] I. Newton, *Mathematical Principles of Natural Philosophy and His System of the World, I: the Motion of Bodies*, UP, Berkeley, Los Angeles, 1974.

[11] R.L. Petritz, "Contributions of Materials Technology to Semiconductor Devices", in: *Proceedings of the IRE*, 1962, pp. 1025–1038.

[12] A. Sarlemijn et al., "Factors in the Development of Modern Physics", in: *Physics in the Making, Essays on developments in the 20th century physics in honour of H.B.G. Casimir on occasion of his 80th birthday*, eds. A. Sarlemijn and M.J. Sparnaay, North-Holland, Amsterdam, 1989, pp. 1-27.

[13] A. Sarlemijn, "Das Galilei-Problem im Kontext der Transistortechnologie", in: *Neue Entwicklungen in der Erkenntnis- und Wissenschaftstheorie*, eds. P. Weingartner and G. Schutz, Hölder–Pichler–Tempsky, Vienna, 1987, pp. 238–244; reprinted in: A. Sarlemijn, *Benaderingswijzen in de technische wetenschappen*, TWIM–Centrum–TUE, Eindhoven, 1986, pp. 75–86.

[14] A. Sarlemijn, "Pendula in Newtonian Mechanics" (forthcoming).

[15] W. Shockley, *Electrons and Holes in Semiconductors with Applications to Transistor Electronics*, 6th printing, Van Nostrand, Princeton, 1956.

[16] W. Shockley, "The Invention of the Transistor — 'An Example of Creative-Failure Methodology'", in: *NBS Special Publication 388*, Monterey California, May 1974, pp. 47–89.

[17] W. Shockley, IEEE Transactions on Electron Devices **23** (1976) 597.

[18] F.S. Goucher, G.L. Pearson, M. Sparks, G.K. Teal and W. Shockley, Phys. Rev. **81** (1951) 637.

[19] G.K. Teal, IEEE Transactions on Electron Devices **23** (1976) 621.

Between Science and Technology
A. Sarlemijn and P. Kroes (Editors)

Chapter 2

Reflections on fundamental physical research in present-day society

L. van Hove

CERN, Geneva

2.1 Introduction

It is an honour and a pleasure to address this Conference dedicated to Henk Casimir's eightieth birthday, and I wish to open by expressing him my warm congratulations and sincere wishes. Henk Casimir is one of the few scientists who combined a very distinguished career in industrial research management with a brilliant research career — to us physicists he is remembered for several outstanding contributions of lasting significance, especially the Casimir operator, the long-range retarded force between atoms, the Casimir effect to which I shall return, all of which are now integral parts of basic knowledge in our discipline.

It is very appropriate, therefore, that this meeting is about Science and Technology. But I found it somewhat more difficult to define the topic which the organizers asked me to address under the elaborate title of "Fundamental Physical Research in Present-Day Society". What I chose to do is to present reflections inspired by the way I tried to interpret the organizer's wishes.

Fundamental physical research is my basic occupation, but I do not think that you expect me to discuss purely scientific matters. My guess is that the key words in the title proposed to me are "in Present-Day Society", and I shall interpret these words to imply two things: I am expected to talk about the relation of fundamental physics to society, and I should consider this relation as it is today, presumably paying attention to differences with earlier times. I shall offer you some reflections along these lines, but I shall try not to fall into the routines of the by now so common science-society discussions. By necessity, some of what I have to say will not be limited to "fundamental physics".

As a brief opening statement, let me say that I share with many the following convictions. Fundamental research is a necessity for mankind, to satisfy our basic need to know as much as possible about ourselves and about the world in which we live, and to help us overcome the problems confronting us collectively and individually. Just as new inventions, new scientific knowledge often leads to beneficial or detrimental consequences and applications. These can only be controlled a posteriori, often because of unforeseeable developments. Such controls and the ensuing restrictions are indispensable, especially in the industrial and military fields. These principles are now widely accepted, and it is reasonable that in our industrially advanced societies research, pure and applied (one should better say "applicable", as often stressed by Casimir) has become beneficiary to considerable financial support and subject to severe public scrunity. I hope that the developing world will follow the same course.

2.2 Growth of science

There are major problems, however. More and more frequently, the frontiers of fundamental research move to such extreme values of the relevant parameters that very large or at least very expensive equipment is needed: large accelerators and telescopes are the classic examples. So far Western Europe has done rather well in this respect by creating joint international facilities like CERN, the European Southern Observatory and other large installations. These institutions are very expensive in construction and exploitation, but the cost is shared among several countries and these joint facilities have very naturally led to forms of research coordination much more effective than what could be reached by purely administrative or budgetary means. This international cooperation has been so successful that it now extends to many other facilities, for example most recently the European Synchroton Radiation Facility project, whose construction in Grenoble started in January 1988. To go further, namely beyond the boundaries of Western Europe, is much more difficult, because a new dimension of intercontinental collaboration involving the superpowers is called for, but the most recent worldwide trends go in positive directions. Another difficulty has also appeared around the large accelerators.

How should one promote the initiative of individual scientists, especially young ones, when the team working on a single experiment often count more than hundred people? Novel ways of organizing research are called for.

All this concerns big science, which grew extremely fast in size and diversity in the last decades. What about small science? Also here growth has been rapid, often with a proliferation of small groups working on the same subjects. Costs are also soaring (think of large computers), but the lack of common large instruments makes efficient coordination more difficult and national entrenchment more frequent. In recent years, much useful work is being done, for example by the European Science Foundation and the European Communities, so that here again there are good reasons for optimism.

Much more severe problems concern the conduct of truly significant fundamental research in the developing world. How could India have a new Raman? Institutions like the International Centre for Theoretical Physics in Trieste, useful as they are, can certainly not provide the solution. Participation of individuals in the advanced projects of fully industrialized countries is again necessary and helpful, but far from sufficient. The problem is so difficult that it would be foolish for me to do more than raise it. I note nevertheless that the developing world would be greatly helped if it had a few Casimirs to guide its efforts in pure and applicable science. We remember the decisive impact of Homi Bhabha in India and the vacuum left after his untimely death.

2.3 Physics-industry relationship

Present-day society also means industry, a very noticeable fact here in Eindhoven, and I shall now make two comments on the relationship of fundamental physical research with industry. In the last decades, big science has given new forms to this relationship, for example the very large industrial contracts of CERN. For the construction of the largest accelerators, the contracts are enormous and one can ask whether such big research projects will lead to a sort of industrial-scientific complex, somewhat analogous to — but of course much smaller and more palatable than — the industrial-military complex so often discussed since President Eisenhower's famous speech of 1961. It is indeed conceivable that some very large scientific instruments be approved and constructed at least partly because of their attractiveness to industry and/or their usefulness for regional development. Many of us would ask: why not if the science is good? Astronomy, high energy physics and space science are valuable domains for such ventures.

My second comment on the physics-industry relationship brings me from very big to very small science. The *Neue Zürcher Zeitung* of 22 November 1988 published an interesting article with, for this conservative newspaper, quite a flashy headline: "Mit Volldampf ins Flüssigkristall-Zeitalter — Japan in der Anwendungstechnik weit voraus". Liquid crystals constitute one of the prettiest subjects in condensed matter physics. They give rise to a multitude of elegant experimental and theoretical studies which should please the purest physicist interested in the order-disorder problematics. Although the first liquid crystal was discovered as early as 1888 by the Austrian botanist F. Reinitzer, it took some eighty years until RCA began the industrial production of liquid crystal displays (LCD), and the technical breakthrough came a few years later with the invention of the twisted nematic cell at Hofmann-Laroche, the industrial company in Basel. Practically within walking distance from Basel begins the Jura region, the homeland of the Swiss watch industry which was undergoing since the fifties a dramatic decline, with deep and painful social consequences. Still, and this is of course the main message of the Swiss journalist, it is not in the Swiss Jura but in Japan that the LCD watches were put on the market starting in 1973. More generally, by converting from traditional watch-making to modern electronics, a region like the Jura could have restored its prosperity and established itself very early as a stronghold of European high technology. For industrial development the interface with small science is more important than with big science, but it seems that quite a few opportunities are missed.

2.4 Nature of scientific knowledge

I now move on to what I personally regard to be the main message of fundamental physical research for the present-day society. It concerns the nature of scientific knowledge and its implications for social behaviour. One of the main tasks of fundamental physical research is to find out about the elementary constituents of matter, their laws of motion and of interaction. Over the last decades, physics has made enormous progress towards this goal, revealing in succession the electron-nucleus structure of the atom, the proton-neutron structure of the nucleus, the quark structure of the proton and neutron. As of today, we describe the inner structure of all matter in terms of quarks and leptons and of their interactions. We describe gravitation by general relativity, the strong interaction by quantum chromodynamics and the electromagnetic and weak interactions by the unified electroweak theory. This so-called "standard model" is amazingly successful in accounting for the available experimental facts in high energy physics, an advance far beyond the most optimistic dreams of twenty years ago. Still, most physicists working in this field are convinced that the standard model cannot be the last word because it suffers from a number of theoretical defects, like too many "basic" constituents, too many free parameters and too little unification of interactions. Previous generations of physicists were confident that the electron was an elementary particle. We are not sure any more.

I believe this situation to be exemplary for the very nature of scientific knowledge: powerful in providing explanations, predictions and applications, but so far always open to revision, calling for improvements and extensions. In an address to the 1955 Autumn meeting of the National Academy of Sciences in Washington, Richard Feynman went even further when he said: "Scientific knowledge is a body of statements of varying degrees of certainty — some most unsure, some nearly sure, but none absolutely sure". This may go a little far — I for one feel absolutely sure of some scientific facts, for example that an atom of the most common isotope of helium is composed of two protons, two neutrons and two electrons. Nevertheless, I regard Feynman's statement as a healthy sign of modesty in one of the top scientists of the postwar generation, so much more satisfactory than the intellectual arrogance of quite a few lesser colleagues. Furthermore, it is often on the basic questions that the lack of certainty is greatest. For example, general relativity is the deepest and most beautiful theory ever constructed, but it is still limited to the classical level and there is no consensus on how to make it compatible with quantum theory. One consequence is that, despite all that is being written about the big bang, nobody knows whether or not there was any space-time singularity in the early universe.

More importantly, science contributes little to the answers which every society tries to give to the truly fundamental questions if the meaning and purpose of life, of morality and religion, and of social progress. Contrary to the broad, mostly worldwide consensus characterizing the state of well established scientific knowledge at any time, the answers to those philosophical and/or religious questions take the forms of convictions and beliefs which differ greatly from group to group, contradict each other and are often invoked to justify conflicts and violence. As I see it, it is here that a lesson of general interest to society can be drawn from the limitations and uncertainties of scientific knowledge, which we recognize so much more clearly today than in the past. If scientific knowledge is always limited and often uncertain, how can one claim unlimited certainty for convictions or beliefs, whether moral, religious, philosophical or political? I consider this to be a message of tolerance and also of pluralism, since there are so many things one cannot know for sure. And no one should be surprised that in countries unfortunate enough to be run dictatorially in the name of this or that set of beliefs, tolerance and pluralism stand out so high among the dreams and demands of the people who are forced to endure the system imposed by the leaders, mostly with deplorable consequences.

2.5 Scientists as advisors of society

But I should get back to my topic, and I do this by noting that, of course, also scientists have convictions and beliefs. Some of them concern scientific research, its consequences and applications, and also here there are differences which can generate conflicts. This can become highly visible in priority discussions for future research projects; the "superconducting supercollider" debate in the USA is a current example. Conflicts occur also, in fact much more frequently, when scientists discuss the social and environmental consequences of applied science (here I mean applied rather than applicable) and of the resulting technologies. I shall comment briefly on the last point.

Nowadays there are countless cases where scientists are called upon to give expert opinions or advice, individually or in committees, to the authorities or to the media, on a multitude of scientific matters of importance to society. In most cases their advice would be of little use if it would rely exclusively on the very small fraction of scientific knowledge which is beyond any reasonable doubt (like the numbers of protons, neutrons and electrons in the helium atom). Should scientists refuse to go beyond? Or, as most do, should they also invoke less certain scientific knowledge? I believe they should do the latter, but — and I regard this "but" to be of extreme importance — with explicit mention of the uncertainties involved. It is evident, however, that this obligation is often neglected, not to speak of those scientific "advisors" who introduce convictions or beliefs under the guise of scientific knowledge.

The practical impossibility in which we have maneuvered ourselves to hold a rational debate on the exploitation of nuclear energy is a sad example of how bad things can become. Equally competent experts driven by opposite convictions give opposite advices and the decision makers cannot be expected to delineate the subtle boundaries between what is known for sure, what is probably true and what is no more than personal belief dressed up in scientific terms. Under such circumstances it may be better to have experts of opposing tendencies prepare reports in the form of possible "scenarios" based on their convictions, and correspondingly biased recommendations. Comparisons between these scenarios and a posteriori debates between the authors may provide a better basis for decisions than the traditional advisory mechanisms. In addition, some of these scenarios may turn out to be interesting science fiction.

2.6 Concluding remarks

Let me then insert a remark on science fiction, a creative activity which is far from insignificant for the science-society relationship. You do not find it only in book form, but also in very respectable scientific journals, for example in *Physical Review Letters* which published in September 1988 a very mathematical article on general relativity entitled "Wormholes, Time Machines and Weak Energy Condition" by the well known physicist K.S. Thorne of Caltech and his collaborators M.S. Morris and U. Yurtsever. With "Time Machines" in the title the taste of science fiction is evident. *New Scientist* showed interest in this work and commented upon it under the headline "Time Machines, Wormholes and the Casimir Effect", which explains why it is quoted here.

A wormhole is a topological deformation of space-time which creates a short passageway between otherwise widely separated regions. Usually the metric of a wormhole is so distorted that no living being could pass it without being destroyed. What Thorne and colleagues do is to try and invent stress-energy tensors in wormholes which could make them traversable. Their proposed solution is based on the Casimir effect and consists in installing in the wormholes appropriately shaped conducting plates which give the desired modifications of the vacuum stress-energy tensor. If this works, one could launch a space expedition in the year Y and arrange it to return to earth well before, perhaps in year Y-50. What a bonus for planning agencies and futurologists; for once they could be right! Clearly, the first spaceship to carry out such an extraordinary mission should be called the Hendrik Casimir Spacetime Explorer.

But, irrespective of the existence of wormholes or time machines, it is not by accident that I conclude on the Casimir effect. It constitutes one of the deepest insights acquired in quantum field theory, initiating what may be called the physics of the vacuum, because it states that the vacuum is modified by boundary conditions in observable ways. The importance of the Casimir effect was only recognized gradually. Now it is understood to be fundamental in physics and cosmology. It illustrates very well the lasting scientific impact of the man we honour today.

Part II: Fundamental Industrial Research

Between Science and Technology
A. Sarlemijn and P. Kroes (Editors)

Chapter 3

Predicting the properties of materials: dream or reality?

M.F.H. Schuurmans
Philips Research Laboratories
Eindhoven

3.1 Why?

As requirements for materials become more varied and tests and experimentation become more time-consuming and expensive, the prediction of the properties of materials will gain in importance[1].

We shall consider this proposition as applied to semiconductors and magnetic materials, which are widely used in electronic components for radio, television, optical communications, and telephone and computer systems. Examples of semiconductor devices are transistors, light-emitting diodes and lasers. Magnetic materials are used in loudspeakers, motors and in magnetic recording.

A wide range of materials can be made today by using special growth techniques. Monolayers of atoms of any kind can be deposited on a substrate. Almost any chemical composition can be made in this way. Examples of such growth techniques are Molecular Beam Epitaxy (MBE) and Metal Organic Chemical Vapour Deposition (MOCVD). In MBE atoms are depos-

[1]The present paper is an extension of a similar paper by M.F.H. Schuurmans, R. Coehoorn, R. Eppenga and P.J. Kelly in Philips Techn. Review **44**, 8/9/10, 1989.

ited on the substrate by evaporation from atomic sources in ultra-high vacuum. In MOCVD atoms are deposited on the substrate from a reactive gas mixture. Both techniques can be used to make a semiconductor laser that emits light at a single, desired wavelength. One important application found in the Compact Disc player is the semiconductor laser, a layered structure containing Al, Ga and As. However, the correspondence between chemical composition and wavelength is not clear at the outset. The usual approach was purely empirical, a matter of trying and seeing whether it worked. But an MBE apparatus might cost more than a million dollars, and because of contamination problems it can only be used for one semiconductor material, i.e. a single chemical compound. While MOCVD is not so expensive and is more versatile, investigating all the possible combinations of materials is still a hopeless task. Predicting the properties of semiconductors from their chemical composition and crystal structure is therefore very important.

Similar arguments apply to the prediction of some of the properties of metals and magnetic materials. Layered magnetic structures of almost any desired chemical composition can be made with the same special growth techniques, but they are expensive and take a long time.

Sometimes new materials with properties that can be determined from the known properties of the constituent materials will be of interest [1]. For the solid-state physicist working on the prediction of material properties this is the simplest case. However, the material is not really new, since the properties of the constituent materials can still be recognized. For "really" new materials another approach is required. The last ten years have seen the development of the branch of computational physics known as *ab initio* calculations. They are called "*ab initio*" calculations ("*ab initio*"means "from the beginning") because the calculations start from the most elementary information available: the atomic number — the number of electrons in the atom involved — and the crystal structure — the positions of the constituent atoms in the material. If any experimental information about the crystal structure of a material is available, it can be used. If the crystal structure is not known, it can be found from the calculations (the crystal structure giving the lowest energy in the ground state is the correct one; this is sometimes called the "energetically most favourable structure"). We shall meet some examples of these calculations later. The atomic number of any atom can be found in the Periodic Table of the Elements (fig. 3.1). The constituents of the semiconductors that are of most interest for applications are found in the part of the Periodic Table shown / hatched. The actual semiconductors of most interest (besides silicon, of course) are GaAs and AlAs, which are used in solid-state lasers and high-speed transistors. Metals of interest are often

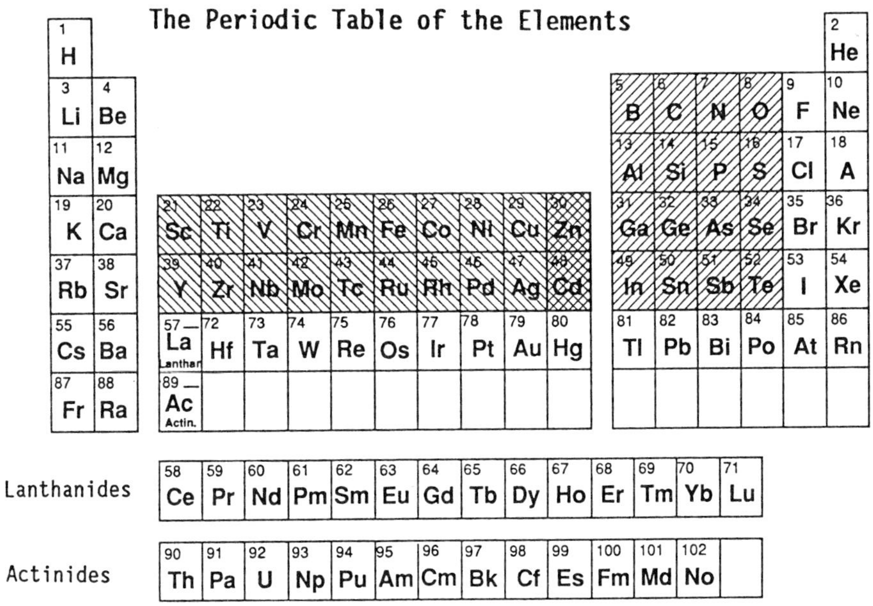

Figure 3.1: *All the input parameters for* ab initio *calculations are given in this table: the Periodic Table of the Elements. The most important semiconductors are formed from elements in the / hatched part of the Table; the most important metals (for use in magnets) are elements from the \ hatched part of the Table or their alloys.*

alloys of elements found in the \ hatched part of the Periodic Table. Elements that are well known as magnetic materials are Fe, Co and Ni. The best permanent magnets are based on the compounds $Nd_2Fe_{14}B$, $SmCo_5$ and Sm_2Co_{17}.

The subject of this article is the *ab initio* calculation of material properties. After evaluating the complexity of this problem we shall discuss the density functional theory [2]. Most *ab initio* calculations for solids are based on this theory. This section as well as the accompanying appendix can be omitted if the reader is not particularly interested in the technical details. We shall then discuss the usefulness of the *ab initio* calculations in predicting the properties of semiconductors and magnetic materials. Finally, we shall draw some conclusions on the extent to which the properties of materials can now be predicted and the success of this approach in theoretical solid-state research.

3.2 How?

So which properties of materials can be predicted and how? To answer this question we first have to see how complicated the problem is. A cube of 1 cm^3 of a material contains about 10^{23} atoms. Each atom consists of a positively charged nucleus and a number of negatively charged electrons. The first problem is that in the material the electrons and the nuclei are all moving. Since nuclei are so much heavier than electrons (about 10 000 times) we can generally assume (but by no means always) that the two motions do not affect one another. So let us assume that the electrons move and the nuclei are stationary. The interaction between the electrons is now the biggest problem. In fact this problem cannot be solved exactly, even for a single atom (unless it is a hydrogen atom). There are a number of useful approximations that can be made, however. If the positions of the nuclei form a regular pattern (the crystal structure), we can reduce the interaction between 10^{23} electrons to the interaction between the electrons that belong to just a few atoms (about 100) — provided, that is, that we are dealing with perfect crystals. In alloys, where the positions of the atoms do not form a regular crystal lattice, the problem is much more complex, and we shall not consider it further.

We have to remember here that the interaction between electrons in a material cannot be described by the laws of classical mechanics. We have to appeal to the laws of quantum mechanics, introduced a.o. by Bohr, Einstein, Heisenberg and Schrödinger in the early 1900s. The interactions between the electrons in a material are described by the Schrödinger equation. From now on we have to bear in mind that the solutions of the Schrödinger equation are states

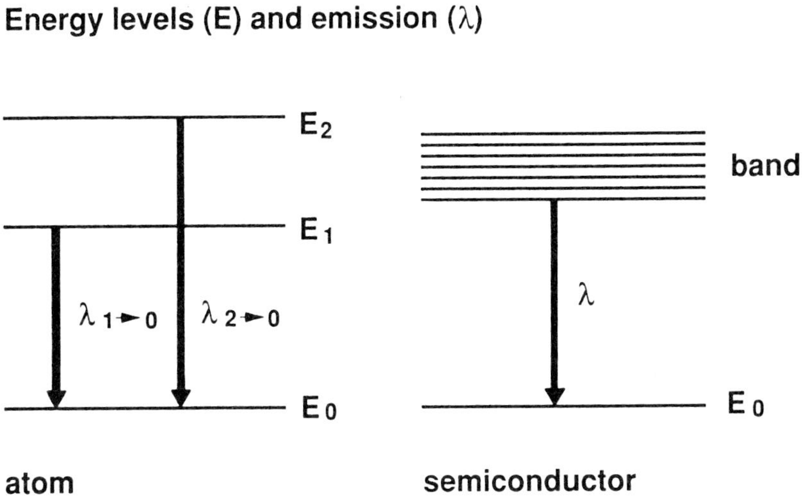

Figure 3.2: *The energy levels for a single atom and a hypothetical semiconductor material. The levels are indicated by horizontal lines and only give the magnitude of the energy. The set of energy levels is called the energy-level diagram. The ground state of a material is the one with the lowest energy, denoted by E_0. The first excited state corresponds to the next higher energy, denoted by E_1. The emission of light, at a wavelength λ, can take place when a material in an excited state returns to the ground state. This is represented symbolically by an arrow. In the semiconductor material energy levels may be grouped in bands. A diagram of this type is also called the band structure*[2].

and the corresponding energies or energy levels. A material can be in one of several states, each corresponding to a different energy. At low temperatures and if it is unaffected by external electric or magnetic fields, a material will be in the state with the lowest energy, the ground state (fig. 3.2). By applying a voltage it is possible to cause the material to go into a state with a higher energy (the energy difference is supplied by the battery). This is an excited state. The emission of light (not necessarily visible) can take place when the material returns to the ground state. The wavelength of the light emitted is precisely determined by the difference between the energies in the excited state and the ground state (conservation of energy). The energy-level diagram, representing the set of energy levels for the ground state and the excited states, is essential to the understanding of the emission of light from a material. Since energy

[2] "Band structures" are most frequently encountered in the context of single-electron energy-band diagrams. Fig. 3.2 shows a diagram for many electrons.

levels often occur in groups or bands, we often speak of band structure rather than energy levels[2].

Even with the fast computers we have today it is not possible to find a solution to the Schrödinger equation that will describe the interactions between 100 electrons. The breakthrough in this field is not primarily due to the increased speed and memory capacity of computers. For the particular class of properties of materials associated with the ground state (crystal structure, lattice parameters, binding energy), the breakthrough has come about because of a theory developed in the sixties: the density functional theory. The *ab initio* calculations based on this theory will be discussed briefly in the next section.

For other properties of materials associated with the excited states, such as the light-emitting properties of a semiconductor, there is as yet no really tractable theory. *Ab initio* calculations of the band structure are then useful, because they give a first indication of the location of the energy levels. A more satisfactory theory [3] has been developed recently, and we shall return to this briefly later.

We can already see that the prediction of the properties of materials is no longer a dream. However, there is much room for improvement, and calculations performed without considering the physical aspects of the problem should be viewed with some reserve. We shall return to this later.

We shall now look at the density functional theory at an elementary level. Readers who are more interested in the answer to the question in the title of this article than in the theoretical physics details may prefer to omit this section.

3.3 The density functional theory

In the early sixties, P. Hohenberg, W. Kohn and L.J. Sham developed the density functional theory, on which most of the present calculations of the properties of materials are based. This theory gives information about a number of the properties of the ground state of an atom, a molecule or, more generally, a piece of material. In principle, the ground state and its corresponding energy can be found by solving the Schrödinger equation, which describes the quantum mechanics of the electrons. However, because of the electron–electron interaction the equation cannot be solved directly. We will give a brief account of the density functional theory; the mathematical formulation is given in the appendix. Hohenberg and Kohn considered a

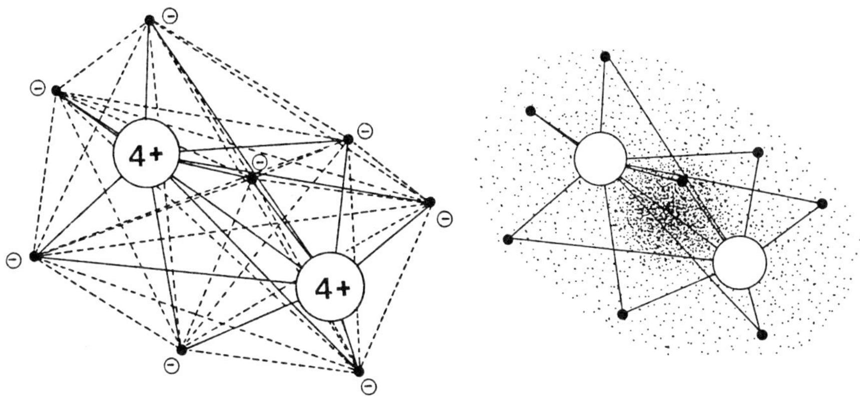

Figure 3.3: *An "artist's impression" of the interaction between ions and electrons in a molecule consisting of two atoms. Interaction between each of the eight electrons and the two ions is indicated by solid lines. The mutual interaction between the electrons is indicated by dashed lines. If the interaction is described by the density functional theory the network of dashed lines disappears and the potential that a hypothetical electron is subjected to is indicated by the dotted background. The variation in the density of the dots corresponds to the variation of the potential.*

function of the electron density[3] that had a minimum corresponding to the energy in the ground state. The function assumes this minimum value for the electron density of the ground state. Some time later Kohn and Sham realised that this minimum formally corresponds to a solution of the Schrödinger equation for a single electron in an effective potential field defined by the many-electron problem. This was a huge step. In simple terms, it means that we started with interacting electrons, which all had to be considered at the same time, whereas now we can consider a single electron influenced by an effective force due to the nuclei and the other electrons. This force can be mathematically described in terms of an effective potential. Fig. 3.3 shows a simplified picture of this description for a molecule consisting of two atoms. It will be clear that density functional theory will make matters even simpler when we have to deal with many interacting electrons. We have to remember that this is a *hypothetical* single-electron problem. This will be particularly important when we come to discuss the band structure.

[3]Electrons move between the nuclei in a material. The electron density corresponds to the probability of finding an electron at a particular position in the material.

There is a price to be paid for this simplification, of course: nobody has so far been able to calculate the exact effective potential, which is a function of the electron density. However, this potential can be approximated by using the known properties of a spatially homogeneous electron gas, and in many cases this so-called Local Density Approximation (LDA) is satisfactory. The approximated potential is still a function of the electron density.

The Schrödinger equation is solved in stages; first the electron density is estimated, then the Schrödinger equation is solved using the approximation for the effective potential based on the estimated electron density. This results in a new value for the electron density. The process of solving the Schrödinger equation and calculating the electron density is repeated until the calculated electron density is the same as the estimated electron density. Now that computers are so much faster and have so much memory capacity this can be done for materials with a complex structure. The essential breakthrough, however, came from the newly acquired physical understanding of the problem.

We have to remember here that the density functional theory does not give the actual ground state. We only find the energy and electron density of a piece of material in the ground state. A theory that gives an approximation to the ground state and also the excited states is the Hartree–Fock theory [2] much used by chemists. This theory has been applied very successfully to atoms and small molecules, but not to pieces of material comprising many atoms. The effective potential in the single-particle Schrödinger equation in this theory does not depend on the electron density, but on a suitably chosen set of functions each of which is a solution of such a single-particle Schrödinger equation. The solution of the single-particle Schrödinger equation in the Hartree–Fock theory is therefore more difficult than in the density functional theory[4].

In the next section we shall discuss a number of applications of the density functional theory. Three of these applications (defects in semiconductors, magnetic materials and the emission of light in semiconductors) relate to research at Philips Research Laboratories, Eindhoven. The other two applications (semiconductors and semiconductor surfaces) relate to work done elsewhere.

---

[4] A version of the Hartree–Fock theory related to the density functional theory is the Hartree–Fock–Slater theory. In Slater's version of the method, the effective potential from the Hartree–Fock theory is approximated by a function of the electron density. Unfortunately, this approximation is often an oversimplification and a correction parameter has to be used.

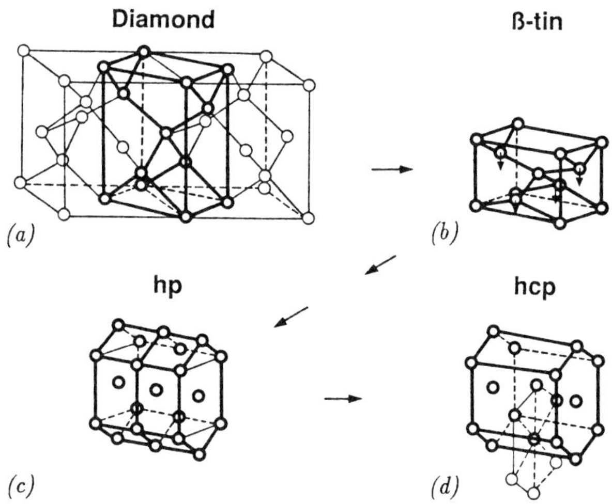

Figure 3.4: *Crystal structures of silicon, for increasing pressure. (a) diamond structure (b) β-tin structure (c) hexagonal primitive crystal structure and (d) hexagonal close-packed crystal structure.*

3.4 Prediction of properties of a material in the ground state

3.4.1 Semiconductors

What can we predict with the density functional theory? Let us first look at silicon, the most important material in the semiconductor industry. Silicon has various possible crystal structures, depending on the applied pressure. At room temperature and atmospheric pressure it has the crystal structure of diamond (see fig. 3.4). Does the density functional theory confirm this? It does indeed. We shall now discuss results mainly obtained by M.L. Cohen and his collaborators at the University of California, Berkeley [4].

The energy of silicon in the ground state can be calculated for different crystal structures and for different distances between the atoms in the lattice. It turns out that the diamond structure gives the lowest energy when the distance between the atoms in the lattice is within 1% of the measured lattice spacing. In addition, the compressibility of the material, that is the change in energy when the material is compressed, can be predicted to within 2% of the measured value. The calculation also shows that above a critical value of the applied isotropic pressure (10 GPa) the diamond structure no longer has the lowest (free) energy [5] (see fig. 3.4a). The arrangement of the atoms with the lowest ground-state energy under these

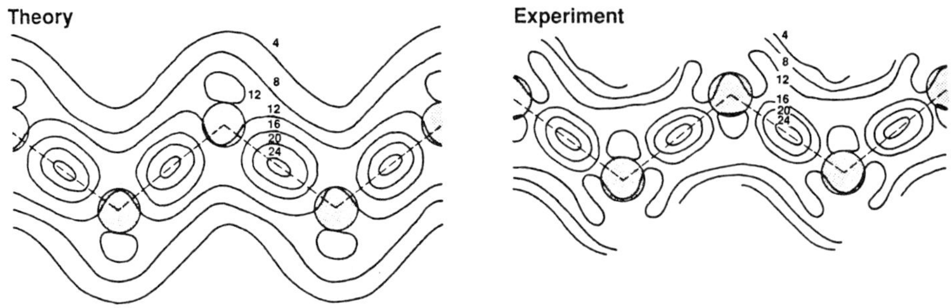

Figure 3.5: *Calculated and measured electron density in a (110) plane in the silicon crystal. Nuclei are indicated by grey circles and lines of constant electron density are shown. The high electron density between the nuclei corresponds to the bonds between the atoms in the crystal.*

conditions can be seen in fig. 3.4b. The new crystal structure is that of β-tin. At still higher pressures we find transitions to the primitive hexagonal structure (fig. 3.4c) at 16 GPa and finally to the hexagonal close-packed structure (fig. 3.4d) at 40 GPa. The predictions indicate that these modifications of silicon under pressure are metals. When silicon is subjected to increasing pressure, experiments do indeed reveal the predicted sequence of crystal structures. Transitions between the different structures occur at the predicted pressure values and under sufficient pressure silicon is indeed a metal. If the temperature is low enough it even becomes a superconducting metal. The calculated electron density of silicon in the ground state in the diamond structure is in excellent agreement with the measured density (see fig. 3.5). Note that these calculations do not involve free parameters. The calculations really are "ab initio". The only input is the atomic number of silicon from the Periodic Table: 14.

3.4.2 Semiconductor surfaces

A good understanding of semiconductor surfaces and interfaces between semiconductors is very important for the continuing development of IC technology. *Ab initio* calculations based on the density functional theory can contribute substantially here. We shall discuss one spectacular example: the (111) surface of silicon.

The atoms at the surface of a semiconductor and those just beneath it are not generally in the positions that might be expected from the periodic arrangement of the atoms in the semiconductor. This behaviour can be explained in the following way. Atoms in the bulk of a semiconductor are bound to each other because the four electrons most weakly bound to

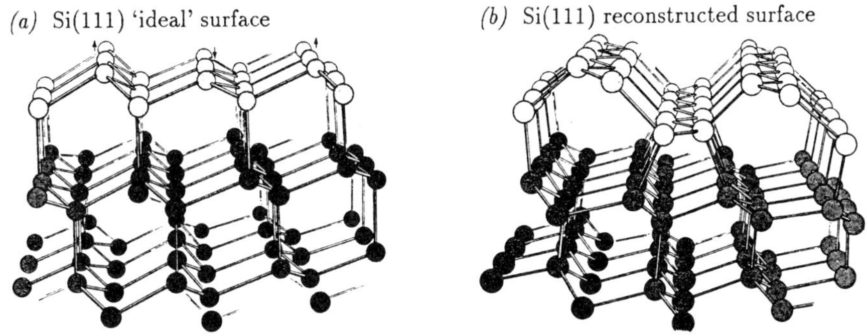

Figure 3.6: *(a) The buckling model of the (111) surface of silicon. As a result of the reorientation of the surface atoms rows of atoms are positioned alternately higher and lower than in the structure in the bulk of the material. (b) π-bonded chain model of the (111) surface of silicon. As a result of the displacement of the surface atoms, alternating pentagonal and heptagonal rings occur. The normal hexagonal structure can be seen in the bulk of the material.*

the atoms are distributed over the four orbitals pointing to the nearest-neighbouring atoms. An atom at the surface, however, has fewer than four neighbours and has thus unbonded or "dangling" orbitals. These dangling orbitals do not point at neighbouring atoms but into the empty space beyond the semiconductor surface. Reordering of the atoms at the surface and those just beneath it, with a partial reorientation of the orbitals, may reduce the distance between the dangling orbitals. The corresponding surface reconstruction, i.e. the new periodic arrangement of the atoms at the surface, results in a lower energy for the surface. The surface reconstruction with the lowest ground-state energy can be found with the aid of density functional theory.

Low-energy electron diffraction (LEED) at the (111) surface of silicon formed by cleaving a crystal shows that the surface has been subjected to a "(2×1)" reconstruction. The notation (2×1) means that the period of the atomic arrangement at the surface in one direction has become twice as large as the period in the bulk of the semiconductor and that it has remained the same in another direction. In the early eighties, Haneman's "buckling" model [6] was the accepted model for the silicon (111) surface; see fig. 3.6a. In this model, rows of atoms are positioned alternately higher and lower than they would be at a non-reconstructed surface. Dangling orbitals do not link up to the orbitals of the nearest-neighbour atoms, but with those of the next-nearest atoms. Some doubt about the correctness of this (2×1) reconstruction of the silicon (111) surface was expressed in 1982, when K.C. Pandey [7] then at IBM Yorktown Heigths showed, with the aid of density functional theory calculations of the ground-state energies, that a model he had invented, the π-bonded-chain model and not the buckling model had the lowest ground-state energy. In Pandey's (2×1) reconstructed model the topology of the atoms has been completely changed; see fig. 3.6b. The dangling bonds now do link up to those of the nearest neighbours. Support for Pandey's model soon appeared from various experiments [8]. If the silicon is heated, the (2×1) reconstructed surface changes into a (7×7) reconstructed surface, now the subject of intense study with the scanning tunnelling microscope [9] ; see also the contribution by J.E. Demuth in this proceedings. The (7×7) reconstruction is unfortunately too complicated for *ab initio* calculations on present-day computers. In a few years, however, the situation may have changed.

3.4.3 Defects in crystal lattices

In the third example of *ab initio* calculations of properties of the ground state we leave materials that have a perfect periodicity (in two directions at the surface, in three directions in the bulk of the material) and we discuss as an example of an impurity (a defect) in a semiconductor some aspects of oxygen in silicon.

The importance of oxygen-related defects in silicon becomes clear when we remember that silicon dioxide, a stable insulating material, is often used in IC technology. As a result of various heat treatments in the manufacturing process, oxygen will always be present in substantial concentrations in silicon, and either on its own or in combination with other impurities gives rise to a large variety of defects. This is also true for nitrogen and carbon. Because these defects may affect the conducting properties of silicon, it is important to understand the physical properties of these defects.

In general, it is not possible to determine experimentally either the position in the lattice of impurity atoms or the positions of the surrounding silicon atoms. Reliable calculations are necessary to complement or confirm the available experimental information. The calculations [10] that we shall now look at briefly were performed by P.J. Kelly at Philips Research Laboratories, and use the most advanced methods available today. They are made in the following way. First of all, the properties of the perfect silicon lattice are calculated. The foreign atom, the oxygen atom, together with the silicon atoms that are displaced as a result of imbedding the oxygen atom, is considered to be a spatially limited disturbance. Since the oxygen atom does not fit into the silicon lattice very well, about ten silicon atoms are displaced significantly. It is because this number is relatively large that it is difficult to find a solution to the problem.

One of the impurity configurations we investigated is an oxygen atom at an interstitial site in the diamond lattice of silicon. Fig. 3.7a shows one of the possible stable configurations in the (110) plane. We would expect the two neighbouring silicon atoms to be pushed aside to make room for the oxygen atom. This does in fact follow from the calculation of the ground-state energy as a function of the Si-O-Si bonding angle and the Si-O distance, but only if the displacements of the silicon atoms further away from the defect are included in the calculations. It turns out that in the configuration with the lowest energy the bond angle and distance are very similar to those found in the insulator SiO_2. Another stable configuration, which for reasons of symmetry has the same energy as that of fig. 3.7a, is shown in fig. 3.7c. How much energy is necessary to move from one configuration to the other? If we assume that the transition takes place by way of the intermediate configuration, fig. 3.7b, the energy barrier to be overcome is the difference in energy between the configurations of fig. 3.7b and 3.7c. The calculated energy barrier is 2.2 eV, in reasonable agreement with the experimental value of 2.56 eV. This is also the activation energy for the diffusion of an oxygen atom through a silicon crystal and it is therefore very important for the "energy housekeeping" of oxygen in silicon. Calculations for defects consisting of several oxygen atoms are at present being carried out.

3.4.4 Magnetic materials

We now come to our last example in this section, magnetic materials. Every electron has an intrinsic angular momentum, the electron spin discovered by the Dutch physicists Goudsmit and Uhlenbeck, and the tiny circular current associated with it makes every electron a little magnet. The electron spin in a magnetic field can have either of two directions, which we shall call spin up and spin down. In non-magnetic materials there are equal numbers of spin-up and spin-down electrons. The magnetic moments of the electrons cancel exactly. In ferro- or ferrimagnetic materials the number of electrons with spin up is not the same as the number with spin down. The magnetic moment is this difference multiplied by the magnetic moment per electron μ_B, the Bohr magneton.

The density functional theory can be extended to describe spin-dependent electron densities so that *ab initio* calculations of the ground-state properties of magnetic materials can be made. For example, calculations have been made to find out whether the spins of a magnetic material

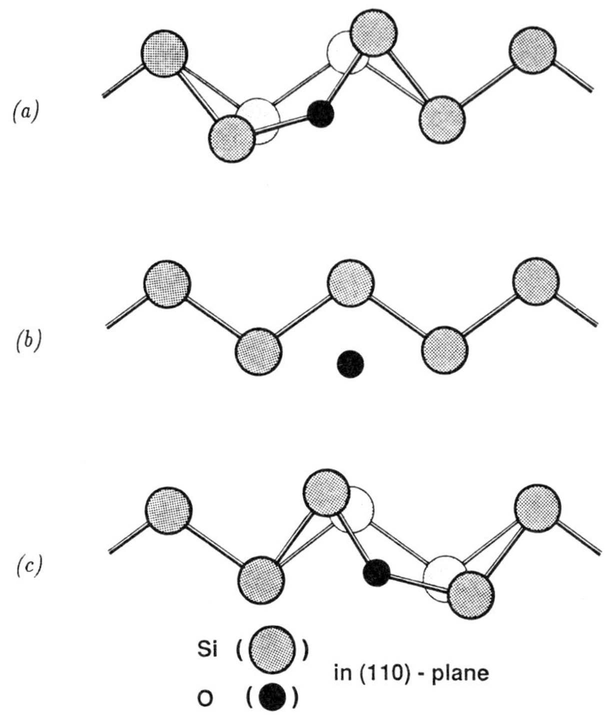

Figure 3.7: *Chain of silicon atoms (grey) in the (110) plane with an oxygen atom (black) at an interstitial position. The oxygen atom does not fit into the silicon crystal lattice: the undisturbed positions of the silicon atoms are indicated by light-grey circles. Configurations (a) and (c) have the same energy, which is lower than that of configuration (b). The calculated energy difference is 2.2 eV. This is an estimate for the activation energy of oxygen diffusion in silicon (the experimental value is 2.56 eV).*

in the ground state are arranged in a ferromagnetic or an antiferromagnetic order [11]. The results are in good agreement with the available experimental data. The magnitude of the magnetic moments can also be compared with experiment. In ferromagnetic materials the magnetic moments of all the atoms point in the same direction. Examples are iron, cobalt and nickel. The calculated values of $2.15\mu_B$ for iron, $1.54\mu_B$ for cobalt and $0.61\mu_B$ for nickel are in reasonable agreement with the experimental results: 2.23, 1.72 and $0.61\mu_B$, respectively. The agreement is even better when the magnetic moment of the electrons in their orbital around the nucleus of the atom (a relativistic effect of about $0.1\mu_B$) is taken into account.

We have already mentioned the permanent-magnet materials $Nd_2Fe_{14}B$, $SmCo_5$ and Sm_2Co_{17} because of their important applications in loudspeakers and motors. The quality of a permanent magnet can be expressed in terms of a number of parameters ($(BH)_{max}$-product, remanent magnetization, coercive force)[5] and depends on the microstructure. Magnetic materials consist of small domains with the magnetizations pointing in different directions. The highest attainable value of the $(BH)_{max}$-product is found to be proportional to the square of the magnetic moment in perfectly crystalline material and can therefore be calculated. The remanent magnetization and the coercive force are more difficult to calculate, because they depend on the domain walls and their displacements under the influence of a magnetic field.

In the search for new materials for permanent magnets attempts are often made to replace some or all of the cobalt atoms in lanthanide-cobalt compounds by iron atoms, which have a slightly larger magnetic moment per atom. *Ab initio* calculations of the magnetic moments of the hypothetical compounds can give some guidance here. We shall now briefly describe some calculations carried out by R. Coehoorn at Philips Research Laboratories [12]. Fig. 3.8 shows the magnetic moment of several compounds of yttrium and iron as a function of the percentage of iron. Yttrium is chemically equivalent to the lanthanides Sm and Nd, but has no magnetic moment; calculations therefore show the contribution of the iron atoms to the magnetic moment. The figure shows experimental and calculated [12] data. The contribution mentioned above from the orbital magnetic moment ($0.1\mu_B$) is not included in the calculations. It should also be noted that the experimental data for the compounds YFe_5 and YFe_{12} were obtained by interpolation and extrapolation, respectively; these compounds are not stable in

[5]The $(BH)_{max}$ product is proportional to the amount of magnetic energy that can be stored in the magnetic field around a permanent magnet. The remanent magnetization is the magnetization that remains in the material after switching off a strong magnetic field. The coercive field is the magnetic field required to demagnetize the material.

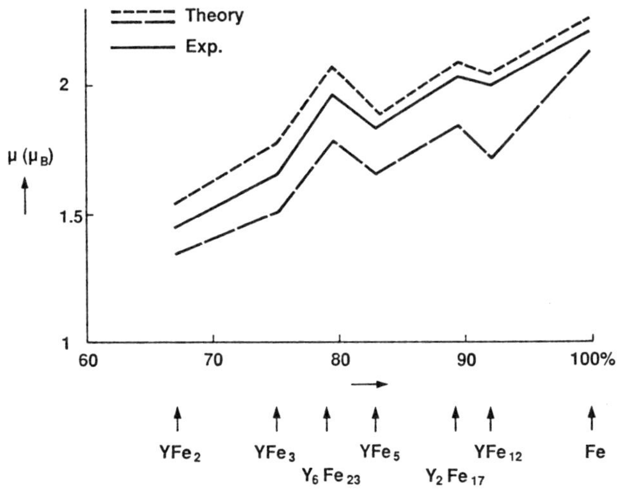

Figure 3.8: *Measured and calculated values of the average magnetic spin moments μ per iron atom for a series of compounds of iron and yttrium. The measured values are indicated by the solid line; the dashed lines refer to the calculated values (small dashes for experimentally determined crystal volume, large dashes for calculated crystal volume).*

the crystal structure assumed. The trend of the experimental results is correctly predicted. The measured values for the magnetic moments lie between the values calculated with the experimental crystal volume and the calculated crystal volume (the difference between the experimental and theoretical curves is generally less than 10%).

In concluding this section we note that density functional theory can also be applied to atoms and molecules. It is found that the bonds between the atoms in a molecule can be described more accurately or more easily with this theory than with existing methods most commonly used by chemists, namely the Hartree–Fock theory. A well-known example is the bond in the molybdenum dimer, Mo_2. The Hartree–Fock theory verified the measured bond lengths, but only after very extensive computer calculations [13]. Density functional theory required far less calculation for the same result [14].

3.5 Prediction of the light emission of materials

When considering the application of materials in LEDs, lasers and television tubes, the electronic structure is extremely important. Can we predict it? Not yet, it seems: density functional theory only gives information about properties of materials in the ground state. This theory makes use of the Schrödinger equation for a single-electron problem, and a band structure can be calculated from this equation. However, this band structure has no real physical significance. In the density functional theory a hypothetical single-electron problem is solved to determine properties of the material in the ground state. States and energies derived for this hypothetical electron in the material are not necessarily related to the states and energies of a real electron. Predictions of the emission wavelength of silicon from this theoretical structure are therefore not correct; the calculated wavelength exceeds the experimental value by 40%. For metals, however, there is often good agreement between the calculated and experimental band structure. Some of the bands in the energy-level diagram for a semiconductor are in fact described very well by the theory. Because of this, some calculations of band structures have been made with the density functional theory, even though there was no proper theoretical or physical basis for using it in this case. Such calculations are called *ab initio* band-structure calculations. Well-known problems [15] such as the monoxides of nickel and manganese show that caution is indicated, however. These materials are good electrical insulators but the theory indicates that they ought to be conductors. Otherwise the properties of these monoxides in the ground state are predicted correctly.

Another aspect is that of the accuracy of the calculations. Comparison of experimental and theoretical band structures and assessment of likely applications for these materials in lasers and LEDs demands an accuracy that the density functional theory cannot provide. In such cases semi-empirical theories, where experimental information is used to meet the desired accuracy, may be of use.

Some calculations of R. Eppenga, S. Colak and the author at Philips Research Laboratories [16,17] of the light emission from superlattices consisting of a periodic structure of alternating thin layers of AlAs and GaAs (see fig. 3.9) may serve as an example. Nowadays very thin layers of these semiconductor materials can be successfully grown with a perfect lattice match by using MBE and MOCVD. The properties of superlattices can be very different from the properties of the semiconductor materials from which the superlattice is constructed. The semiconductor GaAs has a strong infrared emission (λ=870 nm) On the other hand the semiconductor AlAs has a weak emission in the yellowish-green part of the visible spectrum (λ=570 nm). How strong an emission will GaAs/AlAs superlattices give, and at what wavelength? This will depend on the thickness of the GaAs and AlAs layers, of course. If both layers are relatively thick (>40 nm), properties of both GaAs and AlAs will be observed. If the layers are thinner (2–40 nm), the constituent materials will affect one another. Experi-

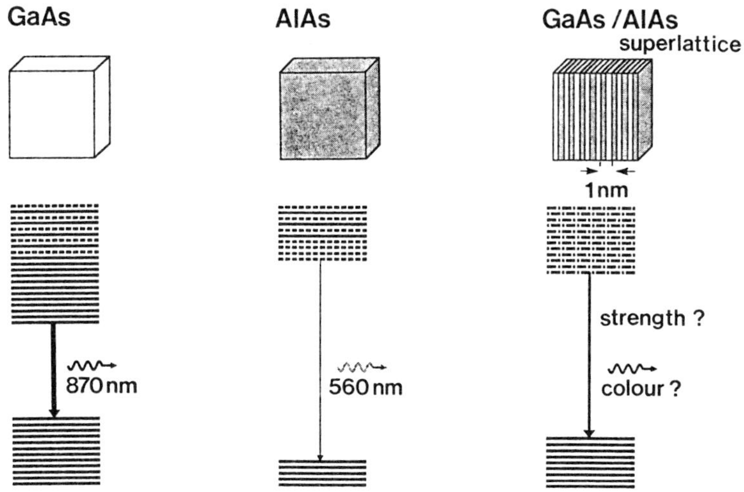

Figure 3.9: *GaAs has a strong emission in the infrared (indicated by a thick arrow). AlAs has a weak emission in the yellowish-green part of the spectrum (indicated by a thin arrow).* Ab initio *calculations help to provide values for the intensity and wavelength of the emission from a superlattice consisting of alternate layers of GaAs and AlAs.*

mentally, a strong emission is found, which originates from the GaAs, but at a wavelength shorter than that normally associated with GaAs and longer than that normally associated with AlAs. Semi-empirical calculations [16] based on the properties of AlAs and GaAs for a superlattice of GaAs and AlAs layers of thickness 4 nm indicate that there should be a strong emission of red light. This is in agreement with the experimental results.

If the layers are even thinner ($<$ 1nm), semi-empirical calculations are no longer permissible, since we now have a truly new material. *Ab initio* band-structure calculations [17] show that weak emission would be expected from a superlattice consisting of GaAs and AlAs layers of thickness about 0.3–1.1 nm. When corrections based on experimental data for AlAs, GaAs and AlGaAs have been made, the calculations show that orange light will be emitted. A comparison with experimental data from perfect superlattices confirms our results.

A theory that gives a correct description of experimentally observed band structures, and therefore the light emission as well, is now gaining ground [3]; see also the appendix. The equations in this theory are very similar to the equations in the density functional theory. This means that density functional theory can be of real benefit in making exact calculations of band structure. Because of its computational complexity the new theory has so far only been used for simple problems [18], but the results are encouraging and further developments are likely. An interesting result concerns the energy of the lowest conduction band of AlAs at the L-point in the Brillouin zone. The "experimental" energy quoted was in fact obtained by extrapolation from the alloy $Al_xGa_{1-x}As$ and is almost certainly wrong by 0.5 eV!

3.6 Dream or reality?

Can we predict the properties of materials correctly yet? The preceding sections show that any answer here must be qualified. For example: predictions of properties of materials in the ground state are realistic, properties associated with excited states should be considered with some reserve, although a new theory definitely shows promise. The theory can only be applied to perfect crystals. Alloys present us with new problems, though there are methods for dealing with them [19]. When we have invented a new material, we often do not know the positions of the constituent atoms very accurately. We can only make further progress if the material has a definite symmetry.

So is the enthusiasm with which *ab initio* calculations have been welcomed in recent years justified? Yes, it is. We can now see the way ahead to making real predictions. Trial calculations of structures consisting of very thin layers of semiconductor or magnetic materials, or combinations of these, which may be of interest to industry now or in future, have already proved useful[6]. The same is true for calculations of bond lengths and angles in large molecules now under way in the petrochemical and pharmaceutical industries. In the preceding sections we have shown that these should be viewed with some reserve. We are still a long way from a situation where service centres can conjure up information about a material from a computer. What we really need now is good team work between computational specialists, physicists and chemists. This is the only way to obtain reliable results. The divide between computational physics and phenomenological physics that some think will be reflected in *ab initio* calculations is therefore undesirable, at least for the time being. And what is more, the question whether calculations should be based on the Hartree–Fock theory or on the density functional theory is misplaced. The information from the two approaches is complementary — certainly for atoms and molecules: energy schemes and states from the Hartree–Fock theory and ground-state properties from both theories.

[6]Such calculations are now being done in the group Advanced Theoretical and Experimental Physics at the Philips Research Laboratories in Eindhoven, the Netherlands. Calculations of the magnetic anisotropy of magnetic multilayers are currently being explored.

It should perhaps be noted here that larger computers alone will not give the answers to the questions we are asking. Much will depend on a better understanding of the physics of interactions between larger numbers of particles. But this in no way removes the necessity for faster computers of greater capacity for future developments in the prediction of properties of materials.

Finally, it should be remarked that the *ab initio* calculation of the properties of materials will be very demanding. It is therefore still useful to develop models [20] for extrapolating properties of materials from experimental data for other materials, since such models can reduce the need for calculation and experiment.

Acknowledgement

My scientific contacts with Casimir have been limited. However, the few I had were extremely inspiring and I am very grateful for them. I still remember vividly Casimir accompanied by V. Weiskopf dropping into my office with the question: "What's new?" At that time, it must have been in 1978, I was working on the phenomenon of superfluorescence in collaboration with Quirin Vrehen, Dick Polder and Hyatt Gibbs. I felt that we had understood the phenomenon quite well. Still I was not able to answer properly to the simple question: "what is the difference with the phenomenon of amplified spontaneous emission?" The net result of that discussion was almost a year of additional work on amplified spontaneous emission and superfluorescence and a far better understanding of the interrelation of these phenomena!

Appendix

The basic elements of the mathematical formulation of the density functional theory will be described for a nonmagnetic system of interacting ions and electrons. The ions are assumed to be at fixed positions (zero temperature theory). Hohenberg and Kohn consider an energy functional E of the electron density $n(\mathbf{r})$ defined by

$$min\,[E_{kin} + E_{ion-ion} + E_{el-ion} + E_{el-el}] \;=\; E[n(\mathbf{r})]\,. \tag{3.A.1}$$

The sum of the kinetic, ion-ion, electron-ion and electron-electron energies is evaluated from the expectation values for the antisymmetric many electron wavefunction Ψ. The minimum is taken over all the many electron wavefunctions Ψ that produce the given electron density

$n(\mathbf{r})$. By variational analysis Hohenberg and Kohn then proved the theorem:

$$E[n(\mathbf{r})] \geq E_{gr} \,, \tag{3.A.2}$$

$$E[n_{gr}] = E_{gr} \,; \tag{3.A.3}$$

i.e the energy functional assumes its minimum value at the ground state density n_{gr}. The minimum value is the ground state energy E_{gr}. Kohn and Sham realized that the minimum can *formally* be found from the self-consistent solution of a one-particle Schrödinger equation when the energy functional is rewritten in the form

$$E[n(\mathbf{r})] = min\,[E_{kin}] + E_{ion-ion} + \int e\, n(\mathbf{r})\, \nu(\mathbf{r})\, d\mathbf{r} + U[n(\mathbf{r})] + E_{xc}[n(\mathbf{r})] \,, \tag{3.A.4}$$

where $\min[E_{kin}]$ is the kinetic energy of a system of non-interacting particles with density $n(\mathbf{r})$, $\nu(\mathbf{r})$ is the potential felt by an electron at $\mathbf{r}$ due to the ions, U describes the electron-electron interaction in an average way (the Hartree approximation), i.e.

$$U = \frac{1}{2} \int e^2\, n(\mathbf{r})\, \frac{n(\mathbf{r}')}{|\mathbf{r} - \mathbf{r}'|}\, d\mathbf{r}'\, d\mathbf{r} \tag{3.A.5}$$

and E_{xc} is by definition the remaining part of E. The trick lies in the formal introduction of a kinetic energy contribution of a non-interacting particle system. By variational calculation one can now easily show that the density $n(\mathbf{r})$ of the ground state can be obtained from the selfconsistent solution of the one-particle equation

$$[-\frac{\hbar^2 \nabla^2}{2m} + \nu(\mathbf{r}) + u(n;\mathbf{r}) + \nu_{xc}(n;\mathbf{r})]\ \Psi_\alpha = E_\alpha \Psi_\alpha \tag{3.A.6}$$

and

$$n(\mathbf{r}) = \sum_{occ} |\Psi_\alpha|^2 \,, \tag{3.A.7}$$

the summation being over the occupied one-electron orbitals. The potential $u(n;\mathbf{r}) = \delta U / \delta n$. The exchange-correlation potential $\nu_{xc}(n;\mathbf{r}) = \delta E_{xc} / \delta n$. The energy of the ground state follows from:

$$E_{gr} \ = \ \sum_{occ} E_\alpha \ - \ U[n_{gr}] \ + \ E_{xc}[n_{gr}] \ - \ \int n_{gr}\, \frac{\delta E_{xc}}{\delta n_{gr}}\, d\mathbf{r} \ + \ E_{ion-ion} \,. \tag{3.A.8}$$

Of course the trick leading to the one-particle Schrodinger equation has put all the difficulties of the problem in the determination of the exchange-correlation energy E_{xc}. One usually resorts to the so-called Local Density Approximation (LDA): the electron gas with a spatially varying density $n(\mathbf{r})$ is locally (at $\mathbf{r}$) approximated by a homogeneous electron gas with the corresponding density $n(\mathbf{r})$. Calculations for the homogeneous electron gas lead to:

$$\nu_{xc} \ = \ \beta(n)\ \nu_x(n) \,, \tag{3.A.9}$$

$$\nu_x = -2e[\frac{3n}{8\pi}]^{1/3} , \quad (3.A.10)$$

$$\beta(n) = 1 + 0.7734x \ \ln(1 + 1/x) , \quad (3.A.11)$$

$$x = \frac{r_s}{21a_B} , \qquad n\frac{4}{3}\pi r_s^3 = 1 , \quad (3.A.12)$$

where $\nu_x(n)$ is the exchange potential of a homogeneous electron gas of density n, $\beta(n)$ accounts for the effect of correlations, a_B is the Bohr radius and r_s measures the inter-electron distance. The justification of the local density approximation is a complicated matter [2] but, most importantly, in most cases where it is applied it works remarkably well. The density functional theory thus produces the energy and the density of the ground state. Moreover, we obtain a spectrum of eigenvalues E_α. However, we have to remember that it is the eigenvalue spectrum of a formally introduced one-electron problem. There is no physical basis to identify this spectrum with that of the single electron excitations. Nevertheless, this identification often works well. In recent years [3] the issue of the single particle excitation energies was addressed from the point of view of the calculation of the single particle Green's function. It turns out that the single particle excitation energies can be obtained from a one-particle Schrodinger equation much like eq. (3.A.6), but with the exchange-correlation potential replaced by a potential which derives from an energy (E_α) dependent self-energy. Solving this equation has produced accurate excitation energies as discussed in the main text. Because of the computational complexity only simple systems can be addressed at present.

References

[1] R. Eppenga and M.F.H. Schuurmans, Philips Tech. Rev. **44** (1988) 137–149.

[2] *Local density approximations in quantum chemistry and solid state physics*, eds. J.P. Dahl and J. Avery, Plenum, New York, 1984; *The inhomogeneous electron gas*, eds. N.H. March and S. Lundquist, Plenum, New York, 1984.

[3] S.G. Louie, in: *Proc. 18th Int. Conf. on the Physics of semiconductors, Vol. 2*, Stockholm, 1986, pp. 1095–1102; R.W. Godby, M. Schluter and L.J. Sham, in: *Proc. 18th Int. Conf. on the Physics of semiconductors, Vol. 2*, Stockholm, 1986, pp. 1103–1106.

[4] M.L. Cohen, in: *Structure and bonding in crystals, Vol. 1*, eds. M. O'Keeffe and A. Navrotsky, Academic Press, New York, 1981, pp. 25–48.

[5] D. Erskine, P.Y. Yu, K.J. Chang and M.L. Cohen, in: *Proc. 18th Int. Conf. on the Physics of semiconductors, Vol. 2*, Stockholm, 1986, pp. 1217–1220.

[6] D. Haneman, Phys. Rev. **121** (1961) 1093–1100.

[7] K.C. Pandey, Phys. Rev. Lett. **49** (1982) 223–226.

[8] R.M. Tromp, L. Smit and J.F. van der Veen, Phys. Rev. Lett. **51** (1983) 1672–1675; P. Chiaradia, A. Cricenti, S. Selci and G. Chiarotti, Phys. Rev. Lett. **52** (1984) 1145–1147; R.M. Feenstra, W.A. Thompson and A.P. Fein, Phys. Rev. Lett. **56** (1986) 608–611.

[9] I.K. Robinson, J. Vac. Sci. & Techn. **A6** (1988) 1966-1970.

[10] P.J. Kelly, in: *Proc. 15th Intern. Conf. on Defects in Semiconductors*, Budapest, 1988.

[11] J. Kubler, A.R. Williams and C.B. Sommers, Phys. Rev. **B28** (1988) 1745–1755.

[12] R. Coehoorn, in: *Proc. ICM*, J. Physique Col. **C8** (1988) 301–302.

[13] M.M. Goodgame and W.A. Goddard, III, Phys. Rev. Lett. **48** (1982) 135–138.

[14] B. Delley, A.J. Freeman and D.E. Ellis, Phys. Rev. Lett. **50** (1983) 488–491; J. Bernholc and N.A.W. Holzwarth, Phys. Rev. Lett. **50** (1983) 1451–1454.

[15] J.H. de Boer and E.J.W. Verwey, Proc. Phys. Soc. (London) **A49**, extra part (31 August 1937) 59–71; J. Zaanen, G.A. Sawatzky and J.W. Allen, Phys. Rev. Lett. **55** (1985) 418–421. The high-T_c superconductors are a recent example where problems are encountered if the density functional band structure is used naively, see e.g. M.F.H. Schuurmans and L.F. Feiner, Ned. T. Natuurk. **A54** (1988) 71–76.

[16] R. Eppenga, M.F.H. Schuurmans and S. Colak, Phys. Rev. **B36** (1987) 1554–1564; For more experimental data see: K.J. Moore, G. Duggan, P. Dawson and C.J. Foxon, Phys. Rev. **B38** (1988) 5535–5542.

[17] R. Eppenga and M.F.H. Schuurmans, Phys. Rev. **B38** (1988) 3541–3544.

[18] R.W. Godby, M. Schluter and L.J. Sham, Phys. Rev. **B35** (1987) 4170–4171.

[19] P. Soven, Phys. Rev. **156** (1967) 809–813.

[20] See for example: A.R. Miedema, Philips Tech. Rev. **38** (1978/79) 257–268.

Between Science and Technology
A. Sarlemijn and P. Kroes (Editors)

Chapter 4

The revolution in surface science: from reciprocal space to real space

J.E. Demuth

IBM Research Division

Thomas J. Watson Laboratory

Yorktown Heights

4.1 Introduction

Surfaces play an important role in a wide range of phenomena that are technological important — from corrosion and catalysis to the fabrication and operation of solid state electronics — not to mention their central role in all biological processes. Despite the undisputed importance of surfaces, this field has not blossomed until the middle of this century. This arises from the fact that physical and chemical processes at surfaces are far more difficult to detect and analyze than in the bulk due to the inherently small amount of material at a surface as well as the reduced symmetry there. Another major factor in the delayed development of surface science has been the need for vacuum technology to allow controlled studies of well-defined surfaces. In its evolution surface science has depended upon the development of vacuum technology and has been driven by the need to understand and harness surfaces in order to solve a variety of problems. Thus, historically surface science has been an interdisciplinary field caught between technical needs and the pure sciences of physics, chemistry and material science. Interestingly a literature search of "surfaces" reveals that in the mid-fifties most "surface" work concerned mathematical surfaces of one sort or another!

4.2 Surface science — "Between science and technology"

Surfaces have nevertheless been important to the development of modern science as we know it today. Consider the photoelectric effects observed by Hertz and others at the turn of the century which were used by Einstein (1905), to help establish the new quantum theory of matter. As an experimentalist Hertz would certainly be captivated if he were to visit one of the synchroton radiation facilities where photoelectron emission studies of all kinds of surfaces are routinely performed using an enormous range of photon energies. In the mid 20s, the Davisson and Germer experiment demonstrated the wave nature of electrons by revealing the diffraction of electrons from a nickel oxide surface. Germer later returned to work in this area and was instrumental in getting Varian Associates to commercialize low energy electron diffraction (LEED) which provides one of the core methods for modern surface science research. It is not surprising that the first issue in 1964 of a new journal entitled *Surface Science* published by North-Holland contains a LEED article by A.U. McRae, a protégé of L.H. Germer. This diffraction method is still flourishing today having been developed over the years to be a primary technique for detailed, quantitative surface structure determination.

As mentioned earlier surface science has been a field of study closely linked to industrial research and the development of vacuum technology. As trivial as it now sounds, light bulbs and vacuum tubes provided the largest impetus to early surface research. At the General Electric Company in 1910, Irving Langmuir worked to improve the electric incandescent light bulb by understanding the influence of vacuum on the life and efficiency of bulbs. He contributed to improving the design of diffusion pumps needed to achieve better vacuums and by 1920 had performed numerous pioneering experiments on the mechanisms of gas adsorption on surfaces. He is also known for his work on the properties of monolayer assemblies and catalytic properties of surfaces and is rightfully considered to be the father of modern surface chemistry. The related work at the Western Electric Company by Davisson and Germer was also directed at understanding and improving the properties of "electron values", i.e. electron scattering and reflection in vacuum tubes. With the new area of solid state physics developing in the 30s, interest in the rectifying properties of new materials grew, ultimately leading to the invention of the transistor. Here, over time, numerous investigators recognized that the variety of photovoltaic and rectifying properties of these junctions were intimately associated with the interface of these materials. Bardeen's proposal of surface states to explain the nature of the rectifying barrier, or Schottky Barrier, in these junctions has to this date continued to stimulate much research. While many workers have pioneered the measurement and characterization of these semiconductor states, a detailed atomic understanding of them has only recently been possible. Generally we can macroscopically characterize the manifestation of these effects, and can propose a reasonable explanation, but we usually lack direct evidence on a microscopic level. As I will discuss later, the revolution occurring in surface science today allows us to directly probe and examine surfaces on an atomic level and promises to resolve such fundamental issues.

The technological development of vacuum pumps and associated equipment (e.g. seals, valves, manipulators, etc.) was not only central to developing reliable light bulbs and vacuum tubes during the early part of the century but was also critical to the field of surface science. Since a monolayer of gas could cover a surface in one second at a pressure of 10^{-6} Torr, one required a much improved vacuum to maintain clean surfaces for detailed studies. This was largely enabled in the 50s by the invention of the sputter-ion pump as well as the ionization gauge to measure lower vacuum pressures. These inventions together with the development of vacuum technology on a commercial basis, by such companies as Varian Associates, allowed ultra high vacuum (UHV) with pressures $\sim 10^{-10}$ Torr to be available by all during the early 60s. The late 60s and 70s witnessed the development of a remarkable variety of methods to characterize and study surfaces using a variety of electron diffraction and electron spectroscopy methods. Here the small penetration and escape depth of low energy electrons provide the inherent surface sensitivity needed for surface studies. These surface sensitive methods, like those applied to study bulk materials, were largely macroscopic measurements which could be used to infer the characteristics of the surface on an atomic level.

In contrast the scanning tunneling microscope developed by IBM scientists Gert Binnig and Heine Rohrer in 1982 is a radically different scientific tool [2]. This allows one to directly probe surface geometric and electronic structure on an atom by atom basis, and to explore atomic level properties impossible to probe directly in the bulk. This is the new revolution in surface science which I will focus on and discuss in this lecture. This elegant new method is also relatively simple and unlike many other surface science methods can be used in non-vacuum environments, even in liquids! I finally note that the STM, like many of its surface science predecessors, was originally developed to probe and analyze small structures in technologically important thin films: yet another development from the interplay of science and technology!

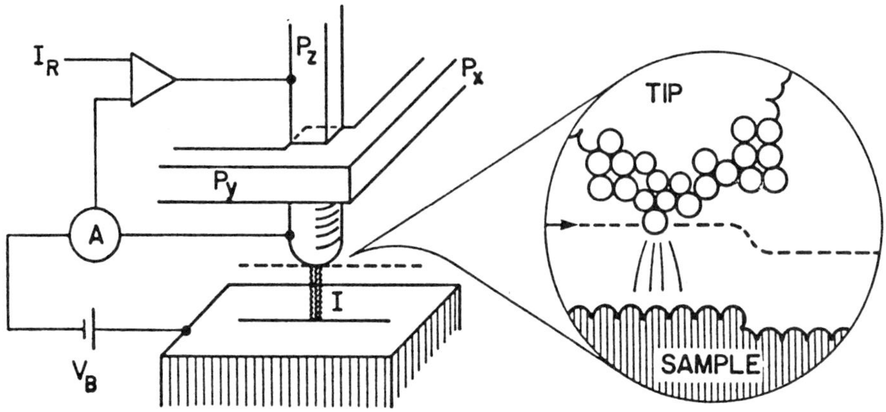

Figure 4.1: *Schematic operating diagram of the Scanning Tunneling Microscope. An enlarged view of sample-tip region is shown to the right.*

4.3 The Scanning Tunneling Microscope

The scanning tunneling microscope (STM) involves positioning an atomically sharp probe tip above a surface as shown schematically in fig. 4.1. The enlarged view on the right shows the "tip" located about 5 Å above the surface. At this distance, the wavefunctions of the sample and tip overlap so by applying a voltage difference between them, anywhere from 1 mV to 2 V, electrons can tunnel from one to the other. This current is monitored as the tip is raster-scanned over the surface by three orthogonal, piezoelectric elements. A feedback mechanism compares the tunnel current to a preset value and produces an offset voltage proportional to the difference. This feedback voltage is then applied to the z-piezo element so as to maintain a constant tunneling current during the x,y-scanning. By monitoring the voltage, and thus the relative position of the z-piezo, one can plot out the contour of the tip as it is raster-scanned along the surface. To zeroth order maintaining a constant tunneling gap makes the tip follow a contour that reflects the topology of the surface.

The high sensitivity and resolution in the z-direction, now typically $\sim$ 0.1 Å, is expected from the exponential dependence of the tunneling current on the separation between sample and tip. A 1 Å change in sample-tip separation produces an order of magnitude change in tunneling current, with typical tunneling currents being $\sim$ 1 nA. The high lateral resolution of about 5 Å was initially very surprising, but later theoretical considerations showed this reasonable for a single or a few atoms on the end of the probe tip. In practice, STM probes are mechanically ground or electrochemically etched and end up with an overall tip radius of 100–1000 Å. The presence of a variety of different asperities, together with the exponential dependence of tunneling, appears to allow a self-selection of the asperity closest to the surface for tunneling.

The most important technical achievement of the STM is the mechanical precision, control and stability available at these small distances. All previous tunneling was performed between materials separated by a thin mechanical layer — usually an oxide. The topografiner, a predecessor to Binnig's scanning instrument, was limited to 1,000 Å resolution due to vibrations and stability problems [3]. Binnig and Rohrer succeeded in vibrationally isolating their new microscope from the outside world as well as developing a method for positioning the sample from relatively large distances to within $\sim$ 5 Å of the sample *without* accidentally crashing the sample into the tip. Today a wide variety of STM's designs are commercially manufactured and available.

Although the STM lends itself well to a wide range of surface studies, it is particularly well suited for studies of semiconductor surfaces. These are the systems I shall concentrate on in the lecture since they are also technologically relevant. As noted earlier, semiconductor surfaces have long attracted interest due to the presence of surface states known to occur since the original work on rectifying contacts. At the surface the broken covalent bonds of the semiconductor are energetically unfavorable and given the right conditions they would like to recombine. Associated with their strongly directional bonding, this re-bonding usually causes a restructuring of the last few surface layers. Since several atoms must reposition themselves when restructuring occurs, some thermal annealing of the sample is usually needed to allow this to occur. Such reconstruction of semiconductor surfaces can be directly probed with the STM as well as the remaining dangling bonds on the surface.

In fig. 4.2a we show a schematic of an idealized model of an unreconstructed, bulk terminated, Si(100) surface and in 4.2b, one model of reconstruction where the dangling bonds of adjacent Si atoms have paired together to form Si dimer pairs. A schematic of the diffraction pattern is shown below where the new periodicity of the dimer pairs produces a larger surface unit cell (shown by the dashed lines) and new reciprocal lattice vectors or fractional order diffraction beams. A detailed multiple-scattering calculation of the intensity variations of these diffraction beams as a function of electron beam energy for various structural models and their comparison to experiment should in principle allow one to determine the detailed geometry of the Si atoms at this surface. While numerous model structures have been considered, none

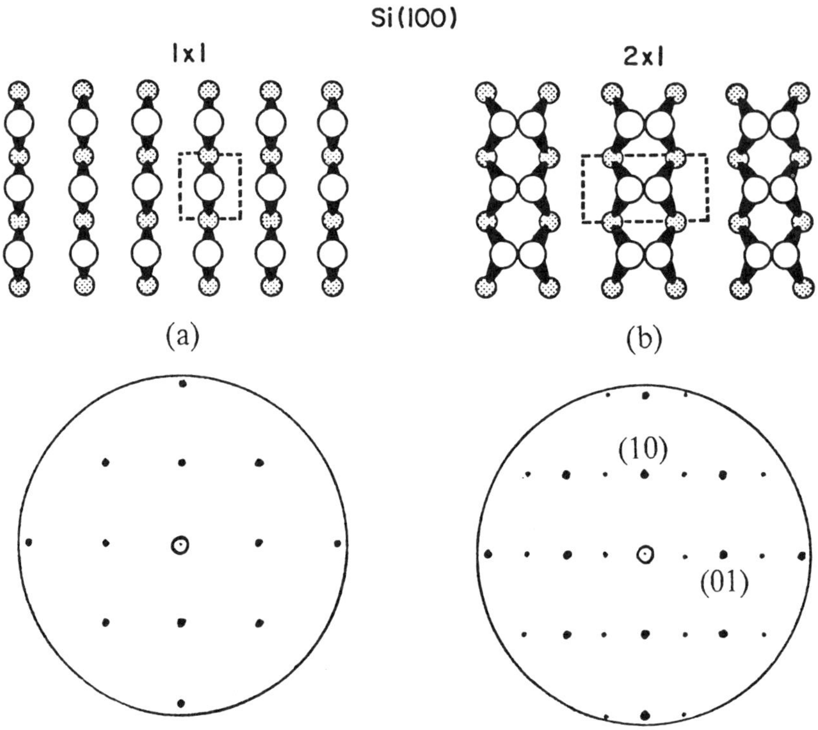

Figure 4.2: *Model of (a) an ideal terminated Si(100) surface without reconstruction and (b) the dimer reconstruction. Below each model is the corresponding reflection electron diffraction pattern where the circular spot in the center corresponds to* $k_{x,y}=0$. *A few of the integral order diffraction beam indices are indicated in (b).*

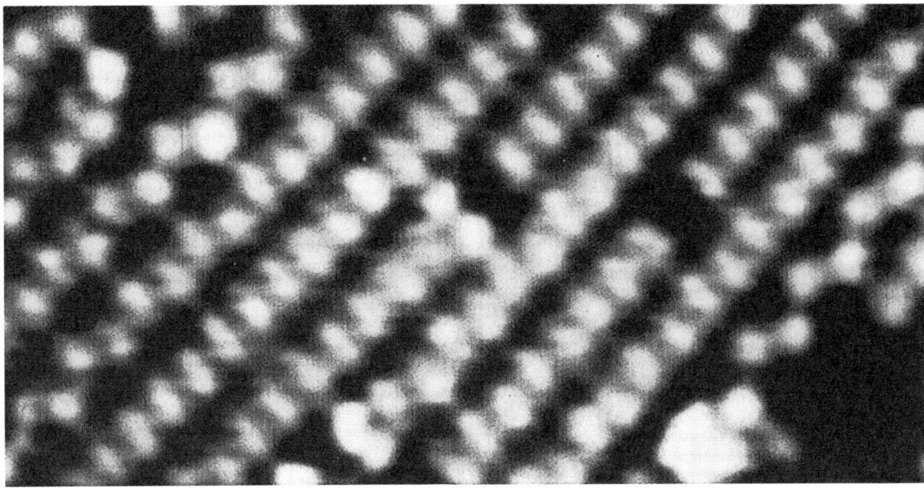

Figure 4.3: *Grey scale image of reconstructed Si(100). The spacing between dimer rows is 7.7 Å.*

provide sufficient agreement to confirm any one model, even the model shown in fig. 4.2b. As will be shown next such an ideal dimer terminated Si(100) surface never exists over the large areas sampled by electron diffraction, but does exist over smaller regions.

One of the first STM images of this reconstructed Si(100) surface formed after annealing an oxide layer off the sample is shown in fig. 4.3 [4]. Here the height scale of this topograph is displayed in a grey scale representation where white is high and black is low. The height range here corresponds to 1 Å. As can be seen the idealized model of Si dimers is not the only structure existing here: numerous missing dimer defects, irregularities and "buckled" dimers also occur. These buckled dimers produce the zig-zag chain like features and arise from a series of tilted dimers whose direction of inclination alternate along the dimer row [4]. A substantially larger area scan is shown in fig. 4.4 and appears to be typical of such thermally annealed clean Si(100) surfaces. The 90° rotation of the dimer rows seen here to occur at mono-atomic terraces of this surface are associated with the tetrahedral bonding of silicon. Namely, each Si atom has four bonds, two that bond "up" to the next layer and two that bond "down" to the layer below. The orientations of the pair of valence bonds projecting out the top of different layers have alternating orientations, rotated 90° in each layer. This layer-dependent alternation leads to two different (orthogonal) dimer pairing directions when these different layers terminate at the surface. The height difference in the terraces shown here corresponds

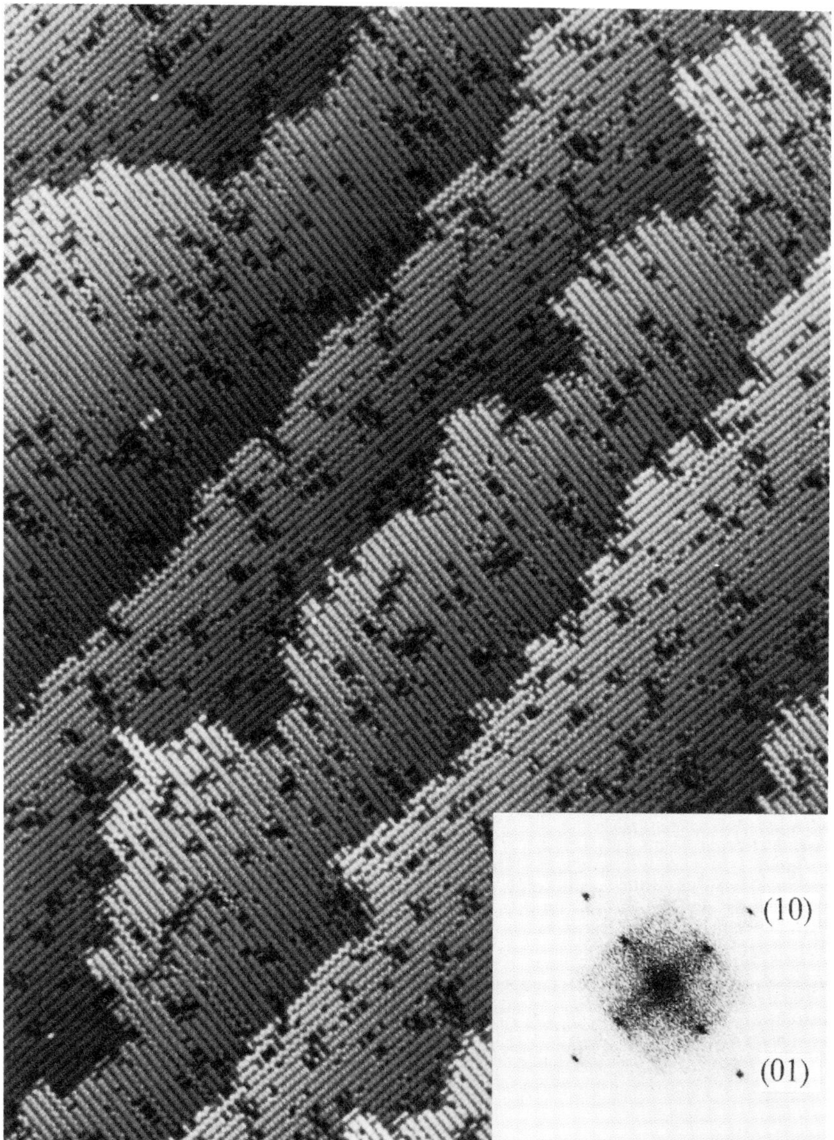

Figure 4.4: *STM image of reconstructed Si(100). The area is 490 Å × 660 Å. The inset shows a two-dimensional Fourier transform of this data and can be compared to the reciprocal space pattern formed by electron diffraction in fig. 4.2b. The inclination of the surface here corresponds to a 0.5° miscut of the sample.*

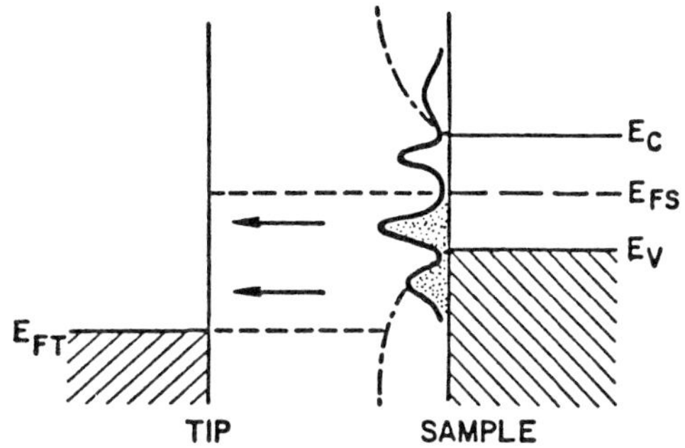

Figure 4.5: *Energy level diagram for tunneling from the filled states of the surface to the empty states of the tip. The surface states that exist on the sample are schematically indicated — whereas the metal tip has a broad continuum of filled and empty states.*

to 1.36 Å the mono-atomic interlayer spacings of Si(100). Shown in the inset in fig. 4.4 is a 2 dimensional Fourier transform of this image which reveals the periodic components of this surface. Due to the two aforementioned orientations of the dimers, "fractional order" periodicities are observed in both directions, not just one as shown in fig. 4.2. This ability to analyze the periodic components of real space STM images provides an excellent compliment to diffraction information.

The atomic scale features seen with the STM actually reflect the charge density contours at the surface and can be used to probe the location of the different surface states that arises. As shown in fig. 4.5 for a given tunneling bias, tunneling occurs from the filled states of the sample to the empty states of the tip (and vice versa for the opposite bias polarity). The tunneling from the filled states of the sample depends on the density of these filled states and for the opposite polarity, the density of the empty states of the sample. Thus, by choosing appropriate bias voltages one can image the filled and empty states of the sample [5]. Fig. 4.6 shows STM images taken at both polarities for 4.6b filled and 4.6c empty surface states of the Si(100) surface. The model in 4.6a corresponds to the same area shown in the STM images. Beneath each STM image is a side view of the calculated charge density contours looking down the dimer row for both the filled and empty states [6].

As can be seen the filled and empty state charge density contours differ primarily due to the node in the empty state wavefunction which produces the strong depletion of charge density between dimer pairs. Overall the image of the filled state reflects the dimer pair.

Spectroscopy can also be performed with the STM by recording I-V spectra at different

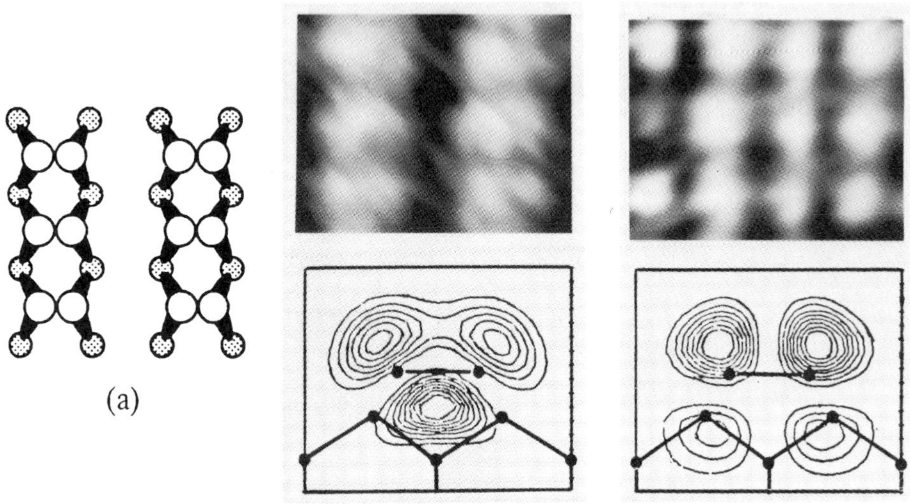

Figure 4.6: *Bias-dependent STM images of Si(100) 2 × 1 compared to an ideal model (a) and theoretical calculations by Ihm et al. [6] for (b) the filled and (c) empty states of the surface.*

points on the surface. To perform such spectroscopy is more difficult than topographic measurement and requires about a factor of 10 greater stability. Recent work by R.J. Hamers and U.K. Koehler [7] illustrates the power of atom resolved spectroscopy on such a highly "defective" surface as Si(100). As shown in fig. 4.7 the I-V spectra near ideal regions show a gap between the filled and empty states — similar but narrower than the bulk band gap. This behavior is expected from theoretical calculations of the dimer model of Si(100) [6]. With the STM however we also find that some of the defects show I-V spectra with a narrower gap and that some defects appear to have even metallic character. This local metallic character represents partially filled surface states at E_F which in essence determines the Fermi level position. Such states have long been proposed to pin the Fermi level.

Another exciting use of the STM, and technologically important, is to understand growth processes on semiconductor surfaces [8]. Shown in fig. 4.8 is the result of less than a monolayer deposition of Si on Si(100) held at a temperature of 370 °C. It is not surprising that the growth structures atop this surface are dimer-like and consist of dimer chains which propagate along the dimer directions. Due to the random nucleation of dimers along the original surface, these dimer rows can be offset from one another by half a dimer, i.e. a single extra row of underlying silicon atoms. As shown in the enlarged inset, such out of registry dimer domains can grow

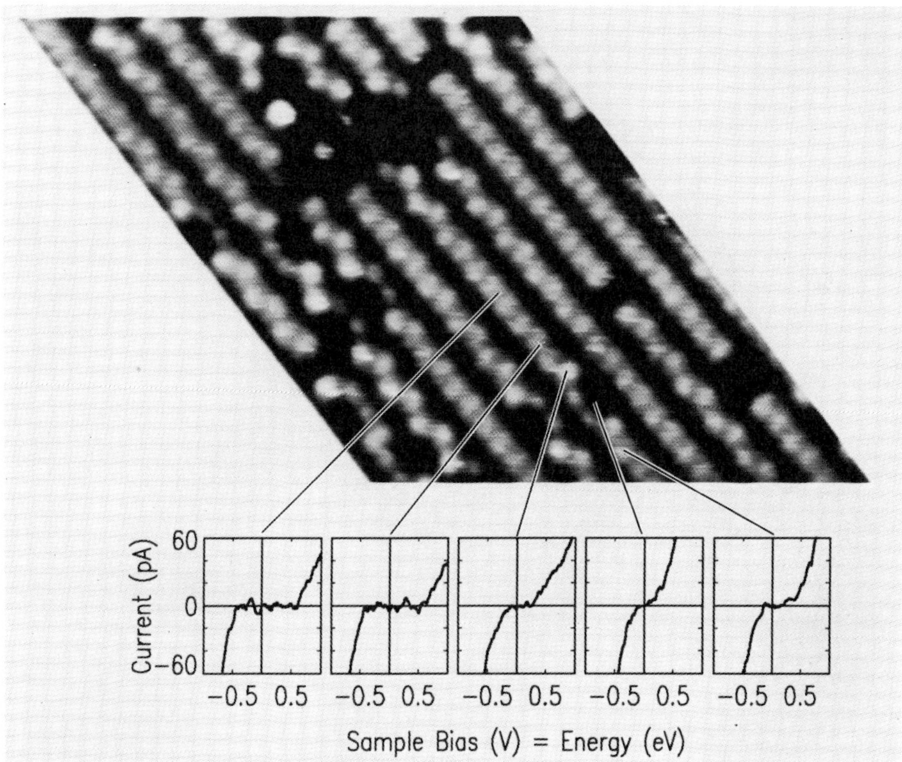

Figure 4.7: *STM images and local I-V spectra taken with atomic resolution at the locations indicated after [7].*

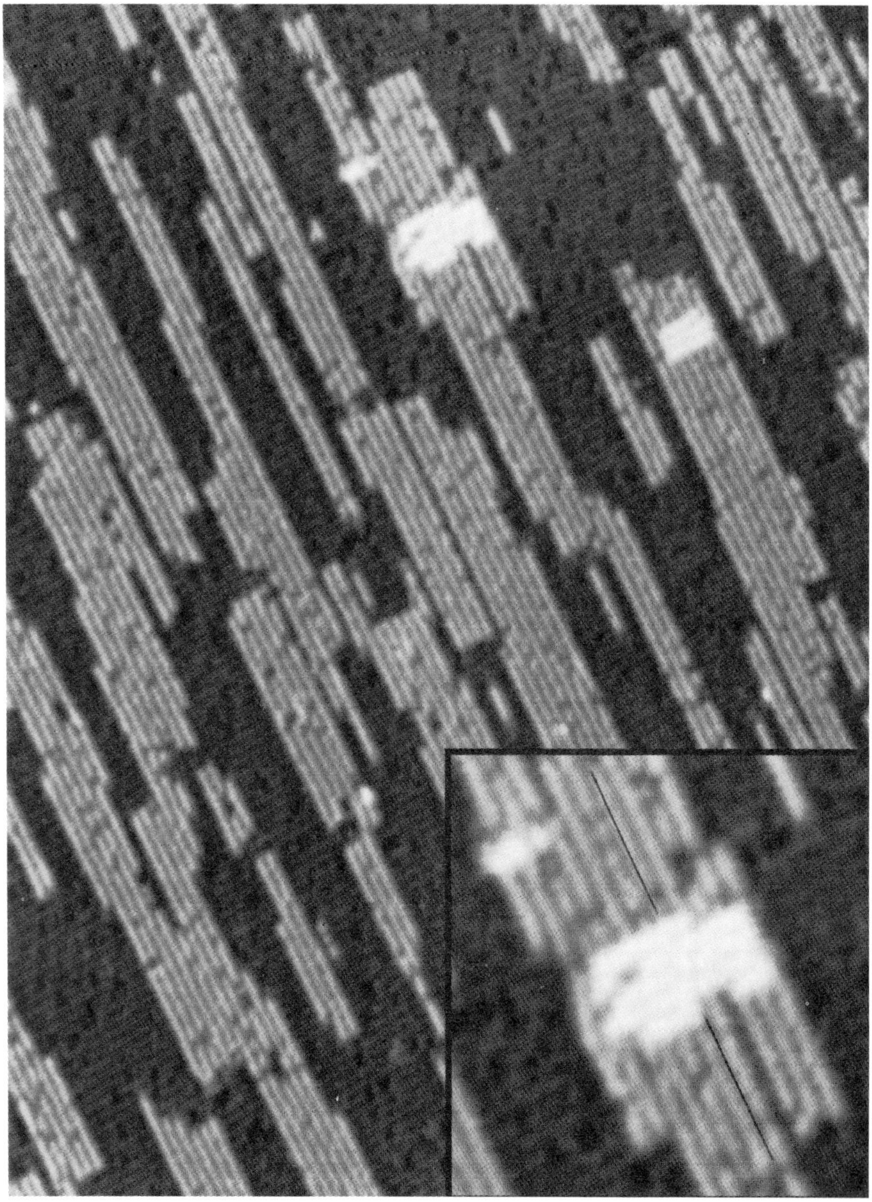

Figure 4.8: *Initial epitaxy for about one half a monolayer of Si on Si(100) at 370 °C. The inset shows a magnified view of an antiphase boundary where second layer growth occurs. The line through this growth region shows the displacement of the dimer row on each side of the boundary.*

together to form an antiphase boundary. These boundaries then tend to form nucleation sites for dimer growth in the next layer. Surprisingly, this initial growth on Si(100) does not "heal" or fill in the defects on the starting surface. However, as can be seen in fig. 4.8 the newly grown dimer structures appear to have fewer imperfections than the substrate. While these observations are just starting to generate new insight for understanding growth, it is clear that surface reconstruction plays an important role in the growth process.

Other surfaces of silicon show different forms of reconstruction. For example, an image from the work of van Loenen et al. [9] on Si(110) is shown in fig. 4.9 and reveals the unusual reconstruction arising in this case. Here strip-like domains occur in the $[11\bar{2}]$ and $[\bar{1}12]$ directions. These strips are terrace-like ridges $\sim$ 20 Å wide whose height is just the Si(110) interlayer spacing. Under higher magnification these ridges appear as zig-zag chains with dimer like features comprising this chain. The complexity of this reconstruction has been a surprise to all and also differs from the last low index face of Si to be discussed next — the (111) surface.

The Si(111) surface is perhaps the most well-known Si surface. Shown in fig. 4.10 is an STM topograph of the reconstructed Si(111)-7×7 surface formed by thermal annealing a Si(111) surface to 1000 °C. It now is well known that the atomic protrusions on this surface correspond to Si atoms bound to three Si atoms of the underlying (111) surface and are referred to as ad-atoms. The surface unit cell is indicated in fig. 4.10. Here missing Si atoms occur in the corner of each unit-cell. There are also Si dimers arising along the sides of the unit cell that cannot be directly seen with the STM which make these boundaries symmetric. To allow ad-atoms to form on both sides of the unit cell about the symmetric dimer boundary, a stacking fault must occur on the last silicon layer on one side of this unit cell [5]. This stacking fault stabilizes the reconstruction and readily occurs at the high temperatures.

Interestingly, this stacking fault poses problems for low temperature epitaxy on Si(111) since it requires more energy to form a stacking fault than that required to form or move ad-atoms around. This can be seen in the topograph of fig. 4.11 obtained after submonolayer deposition of Si on a Si(111)-7×7 surface held at 375 °C [8].

In contrast to Si(100), epitaxy on Si(111)-7×7 forms well-defined clusters of Si ad-atoms. However, at these temperatures the ad-atoms form small clusters of usually incomplete unit cells of the reconstructed surface. Further, the Si atoms that incorporate into the step edge are also very disorganized and do not form or extend the original surface unit cells. Both of these features are due to the inability to form a stacking fault at low these temperatures. At

Figure 4.9: *STM image of the reconstructed Si(110) surface. The image area is 1,700 Å × 1,700 Å. (Courtesy E. van Loenen, Philips Research Labs, Eindhoven.)*

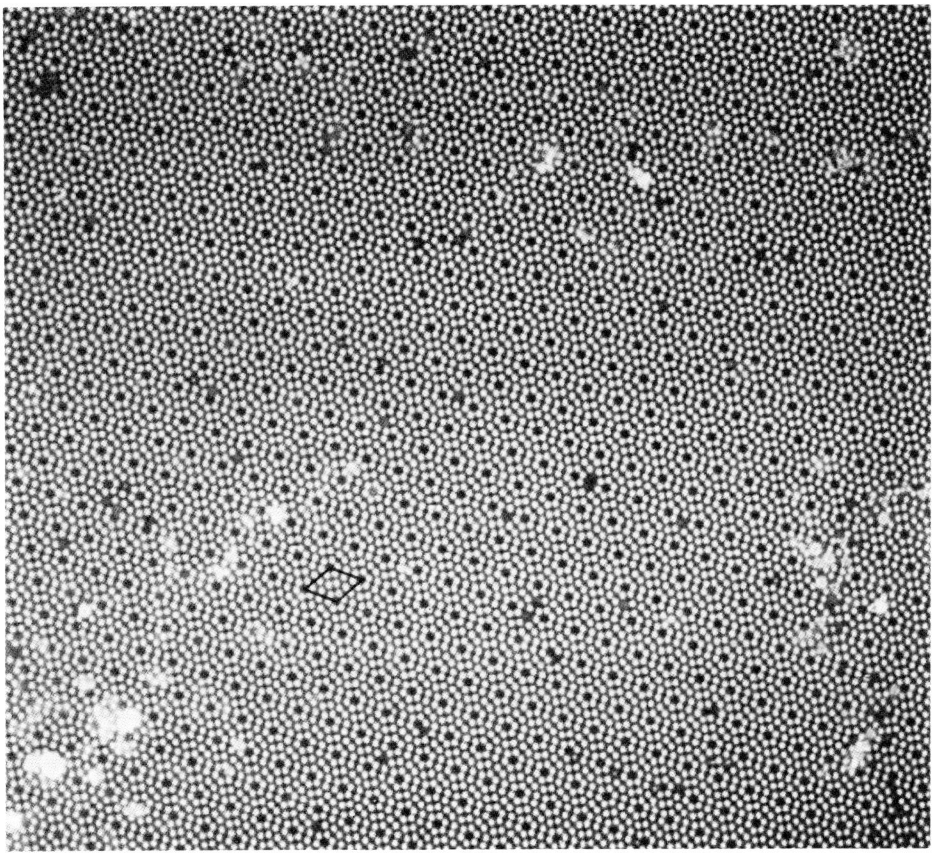

Figure 4.10: *STM image of the reconstructed Si(111) surface. This image area is 700 Å × 750 Å where the unit cell shown is 26.88 Å on each side — exactly 7 times larger than the substrate unit cell.*

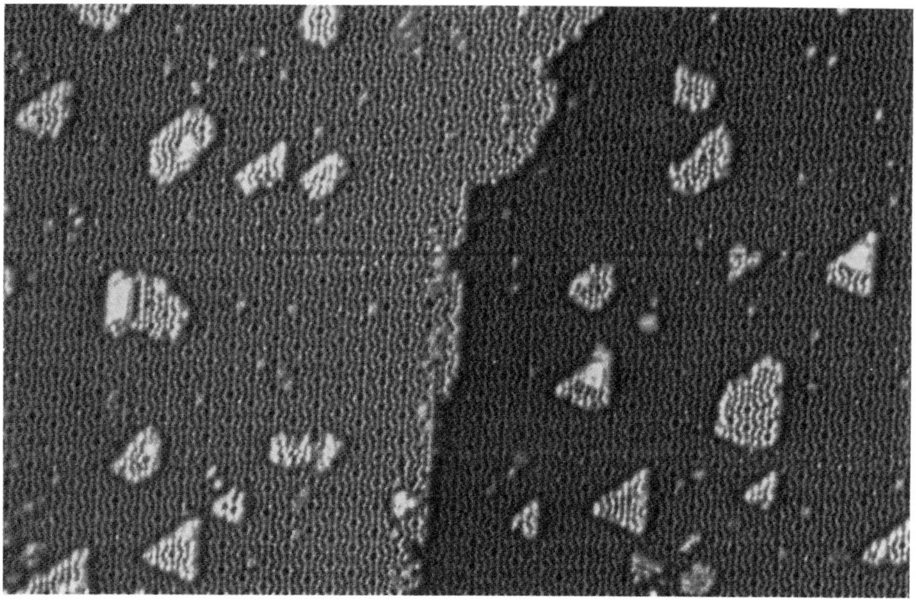

Figure 4.11: *Initial epitaxy of Si on Si (111) at 375 °C.*

higher temperatures the stacking faults can form and then allow large areas of reconstructed Si to form almost perfect island structures.

From the examples discussed it is clear that the STM can provide a wide range of unique atomic level information about surfaces. The features we can now observe are much more complicated than we have ever imagined! From examining the larger area images, we also find a wide range of behavior and variations, in a sense expected from nature. Although the examples presented here are characteristic of recent UHV studies, the STM is also being applied to study a variety of technical surfaces in air, such as magnetic recording media, compact discs or mirror surfaces as well as biological systems and crystalline electrodes in electrochemical solutions to mention but a few. Undoubtedly, this new technology — the STM — has and will continue to provide new scientific insight about the atomic details of surfaces not possible before.

Acknowledgement

The author wishes to acknowledge the numerous contributions of his collaborators, R.J. Hamers, U.K. Koehler and R.M. Tromp, as well as the Office of Naval Research for partial support of this work. The astute secretarial skills of R. LaCorata are gratefully acknowledged.

References

[1] A.U. McRae and L.H. Germer, Surface Science **1** (1964) 319.

[2] G. Binnig and H. Rohrer, Helv. Phys. **Act 55** (1982) 726.

[3] R. Young, J. Ward and F. Scire, Rev. of Instrum. **47** (1976) 1303.

[4] R.M. Tromp, R.J. Hamers and J.E. Demuth, Phys. Rev. Lett. **55** (1985) 1302.

[5] R.M. Tromp, R.J. Hamers and J.E. Demuth, Science, **234** (1986) 304.

[6] J. Ihm, M.L. Cohen and D.J. Chadi, Phys. Rev. **B21** (1980) 4592.

[7] R.J. Hamers and U.K. Koehler, Jour. Vacuum Science and Technology **A7** (1989) 2854.

[8] U.K. Koehler, J.E. Demuth and R.J. Hamers, to be published.

[9] E.J. van Loenen, D. Dijkkamp and A.J. Hoeven, J. of Microscopy **152** (1988) 487.

Part III: Foundations of Physics

Between Science and Technology
A. Sarlemijn and P. Kroes (Editors)

Chapter 5

Gravitation and elementary particle physics

G. 't Hooft

Institute for Theoretical Physics
Utrecht

5.1 Introduction

One of the most challenging problems in theoretical physics is to reconcile quantum mechanics with general relativity. Both theories seem to follow from fundamental principles but they are very different. The problem is that the principles are, to some extent, found to be in conflict with each other. We do not know whether the unknowns are overdetermined or undetermined by the equations we have; so far they certainly are ill-determined. And therefore it is far more likely that the equations we use as input are only partly or approximately correct. Churning them will not help; we'll have to change them. This is where, temporarily, we sometimes have to abandon highbrow mathematics and search for new improved physical principles instead. We may find results that disagree with others, not because a mistake was made somewhere, but because different equations were used.

The search for new physical principles and equations is often more delicate and uncertain than journeys into the wonderful world of mathematics, which might explain why most publications on the basic forces focus on their mathematical aspects. Of course any "new" set of equations should also be mathematically sound, so the above should not be used as an excuse to avoid mathematics, but serve as a warning that at all stages of a mathematical excursion we should keep the physical interpretations in mind, and ask ourselves how sure we are that the equations are still valid here. It is questions of this sort which I focus on in this lecture.

Gravitation is the weakest and the strongest force known in physics. When considered as a force between two single electrons, it is nearly 43 orders of magnitude weaker than the electro-magnetic force. But gravity works collectively: when an amount of matter, somewhat more than the mass of our Sun, is allowed to cool and compress under its own weight, then sooner or later a complete collapse has to take place. No other force can then overcome the gravitational one. The process of collapse can be computed using well-established laws of physics, and few physicists doubt on the final outcome: a black hole [1].

Black holes form an extremely interesting theoretical laboratory. A quite surprising result was found by S. Hawking [2] in 1975. Applying quantum field theory to the region surrounding a black hole he discovered that a black hole must radiate matter of all sorts, behaving like an ideal radiating black body with a temperature T given by

$$k_B T = \frac{1}{8\pi M}\,, \tag{5.1.1}$$

where k_B is Boltzmann's constant and M the black hole mass, in units where $\hbar = c = \kappa = 1$ (κ is Newton's gravitational constant).

It is often forgotten that this result did require certain assumptions concerning the correctness of the interpretation of quantum mechanical wave functions near the horizon of a black hole; one *could* consider the possibility that quantum mechanical probabilities are different near black holes, in particular for "observers" who actually fall into the hole and will not be able, for fundamental reasons, to communicate their findings to the outside world. One such theory gives a value for the black hole temperature that differs by a factor of two from the usual one [3].

But that black holes radiate, with some temperature, loosing energy, hence mass, is very likely. The mass loss per unit of time is estimated to be

$$-\frac{\mathrm{d}M}{\mathrm{d}t} = C_1 \cdot Area \cdot T^4 = C_2 \cdot M^2 \cdot M^{-4} = C_2 M^{-2}, \tag{5.1.2}$$

where $C_{1,2}$ are constants or just slightly depend on T. So one expects the mass to decrease as

$$M(t) \cong C_3 (t_0 - t)^{1/3}. \tag{5.1.3}$$

What will the final stages be like? Is there a lower limit for the black hole mass? If so this can only be when either quantum field theory or general relativity, or both, cease to be valid. It is difficult to object against the fundamental principles of these theories as long as the gravitational force can be quantized perturbatively, and this is the case when all distance scales used are considerably larger than

$$\sqrt{\frac{\kappa\hbar}{c^3}} = 1.6 \cdot 10^{-33} \text{ cm}, \tag{5.1.4}$$

called the *Planck length*. At this length scale we have energies of the order

$$\sqrt{\frac{\hbar c^5}{\kappa}} = 1.22 \cdot 10^{22} \text{ MeV}, \tag{5.1.5}$$

the Planck energy. One could expect that the lightest black hole has a mass of this order, but at this energy black holes may well become indistinguishable from elementary particles. After all, elementary particles with such a mass would be surrounded by the same gravitational field, so they also can be seen as "collapsed objects", and it may well be that all particles at such energies are unstable, emitting radiation much like black holes do while they decay.

Plausible as this picture may seem (and indeed we will adopt it), there are problems with it. It suggests namely that black holes should form a discrete spectrum just like elementary particles, to be characterized by quantum numbers. Also larger black holes could then, at least in principle, be characterized by discrete quantum numbers, and this is not what one gets applying Hawking's techniques.

Either something is wrong with our conventional picture of what matter is like at the Planck scale, or something is missing in the theory leading to Hawking radiation, or both.

We will briefly explain the phenomenon of Hawking radiation, the assumptions that went into the derivation of this, as yet entirely hypothetical effect, and how one could change one's starting points if one wants to obtain results that are easier to reconcile with other ideas in elementary particle physics. We'll see that some very fundamental aspects of the interpretation of quantum mechanics are touched upon.

5.2 The black hole

A black hole is a solution of the classical Einstein equations for the space-time metric, and the equation of state for matter, such that space-time can be seen to be divided in two regions, "region *I*" and "region *III*" (called that way for no particularly good reason), which are defined as follows:

(i) All points in *I* can be connected to the outside world by a timelike geodesic directed into the future, and

(ii) None of the points in *III* can be connected that way to the outside world.

We have to explain what is meant by the "outside world". This concept only means something if we have a non-compact asymptotically flat region surrounding the "black hole" solution at $t \to \infty$. In contrast, region *III* can at all times be enclosed inside a closed surface with an area never exceeding a certain number Σ. For a black hole with an infinite lifetime the question whether or not a signal will be able to escape from region *III* will be unambiguous.

The (lightlike) surface dividing the two regions is called the future event horizon, or horizon for short[1]. Within the horizon, in region *III*, all light cones are directed inwards, so that no information can escape. In the center we usually have some sort of *singularity.* The singularity is essentially a divergence, where presumably our known laws of physics break down. Remarkably however, the singularity will not play a significant role in our discussion. This is because it is well shielded from the observable world; we don't see it, so its precise nature is irrelevant. In this lecture I'll mainly focus on the horizon.

As can be seen from fig. 5.1, a black hole is expected to rapidly settle for a stable, stationary configuration. In a particular coordinate frame the space-time metric can then be written as

$$\begin{aligned} \mathrm{d}s^2 &= -\left(1-\frac{2M}{r}\right)\mathrm{d}t^2 + \left(1-\frac{2M}{r}\right)^{-1}\mathrm{d}r^2 + r^2\mathrm{d}\Omega^2\,; \\ \mathrm{d}\Omega^2 &= \mathrm{d}\vartheta^2 + \sin^2\vartheta\,\mathrm{d}\varphi^2. \end{aligned} \tag{5.2.1}$$

where M is the mass of the black hole in natural units. This we refer to as the Schwarzschild solution.

In this representation there may seem to be a singularity at $r = 2M$, but this is a coordinate artefact. There is a better set of coordinates, called "Kruskal coordinates", in which we see that (5.2.1) in fact describes part of a larger space-time [4]. Replace r and t by x and y, defined by

$$\begin{aligned} xy &= (1-\frac{r}{2M})e^{r/2M}\,; \\ \frac{x}{y} &= -e^{t/2M}. \end{aligned} \tag{5.2.2}$$

Then in terms of x, y, ϑ and φ we have

$$\mathrm{d}s^2 = -2\,A(x,y)\,\mathrm{d}x\mathrm{d}y + r^2\,\mathrm{d}\Omega^2, \tag{5.2.3}$$

[1]The fact that black holes have only a finite lifetime affects our definition of the horizon: it makes the horizon somewhat "fuzzy". However, the uncertainty this induces to the location of the horizon will be far less than other fluctuations that we will discuss.

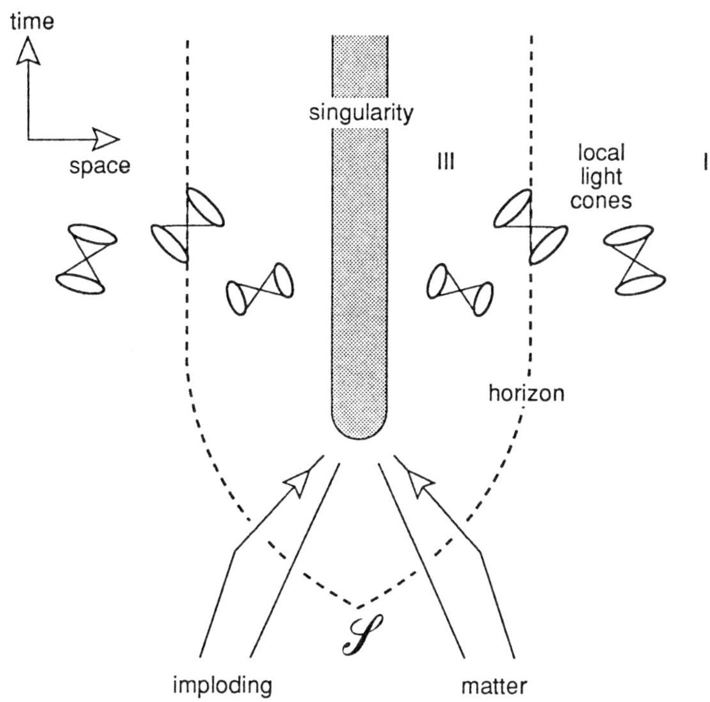

Figure 5.1: *The formation of a black hole.*

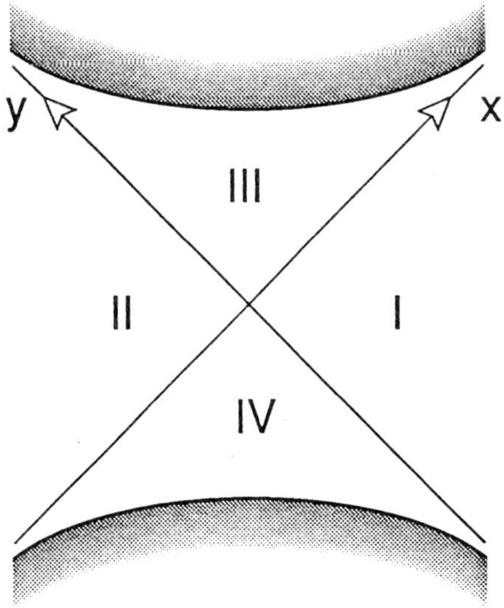

Figure 5.2: *The Kruskal coordinates.*

with $d\Omega^2$ as in eq. (5.2.1) and we find

$$A(x,y) = \frac{16M^3}{r} e^{-r/2M}, \tag{5.2.4}$$

which is not singular at $r = 2M$. Notice now that for every r and t we have in general *two* solutions for x and y, differing from each other by a sign. This implies that our universe is smoothly connected to another, equal, universe, and we obtain regions *II* and *IV*, see fig. 5.2.

If the Kruskal coordinates are used to describe the black hole of fig. 5.1 then the regions *II* and *IV* in there are unphysical. This is because the infalling matter that created the black hole was ignored. Region *IV* would be at the infinite past only if the black hole is "eternal": it existed already before the universe was there[2]. Region *II* can only be reached from region *I* by travelling faster than the speed of light. In realistic black holes (the ones formed by a collapsing object) regions *II* and *IV* are replaced by flat space. On the y-axis ($x = 0$) one then has the imploding material which gives a right-hand side to Einstein's equations such that this hybrid configuration (regions *I* and *III* glued onto flat space) is a solution.

[2]More about the "eternal" black hole in section 5.7.

5.3 Field Theory in Rindler Space

Let us now concentrate on the region $r \cong 2M$. Write

$$x = t_1 + z_1 \; ; \qquad y = t_1 - z_1 \; ; \qquad \frac{r}{2M} - 1 = \frac{\zeta^2}{\rho} \; ; \qquad \frac{t}{4M} = \tau . \qquad (5.3.1)$$

We look at a small angular region so that ϑ and φ can be replaced by two transverse coordinates $\tilde{x}_1 = (x_1, y_1)$. Close to the horizon ($r = 2M$ or $\zeta \simeq 0$) we have

$$z_1 = \zeta \cosh \tau \; ; \qquad t_1 = \zeta \sinh \tau . \qquad (5.3.2)$$

We call $\tilde{x}_1$, z_1 and t_1 the locally regular coordinates. ζ and τ are the Rindler coordinates [5]. We see that ζ and τ are directly related to the Schwarzschild coordinates r and t. In terms of the regular coordinates space-time is not at all singular near the horizon.

Note now that a shift in the τ parameter (or equivalently in the Schwarzschild time t) corresponds to a Lorentz boost in x_1-space. All equations of physics are invariant under such boosts and therefore invariant under shifts in Rindler (or Schwarzschild) time.

A field theory — any local field theory — can be defined by giving the Hamiltonian as the integral of a Hamilton density over 3-space. In ordinary flat coordinates this is

$$H_{reg} = \int \mathcal{H}(\mathrm{x}) \mathrm{d}^3\mathrm{x} \, . \qquad (5.3.3)$$

The Hamiltonian is the operator that generates shifts in time t_1. Now at $\tau = 0$ an infinitesimal shift $\delta\tau$ in τ corresponds to a shift

$$\delta t_1 = \zeta \delta \tau , \qquad (5.3.4)$$

in t_1, as one can easily see from (5.3.2). Thus, the generator for an infinitesimal translation in τ is

$$H_{Rin} = \int \mathcal{H}(\mathrm{x}) \, \mathrm{z} \, \mathrm{d}^3\mathrm{x} = H_I - H_{II} \, , \qquad (5.3.5)$$

with

$$H_I = \int_{z>0} H(\mathrm{x}) \mathrm{z} \mathrm{d}^3\mathrm{x} \; ; \qquad H_{II} = \int_{z<0} H(\mathrm{x}) |\mathrm{z}| \mathrm{d}^3\mathrm{x} \, . \qquad (5.3.6)$$

We recognise (5.3.5) as the generator of a Lorentz transformation.

In most field theories it is not difficult to check that

$$[H_I, H_{II}] = 0 \, , \qquad (5.3.7)$$

because the integrands in (5.3.6) vanish at $\mathrm{z} = 0$. Indeed, one expects (5.3.7) because no signal can be transmitted between regions I and II.

To describe physical processes at the visible part of a black hole one needs only H_I. Would particles described by H_I alone "bounce" against the horizon? To see what happens it is instructive to consider a solution of a wave equation close to the horizon. Writing

$$\zeta = e^{\sigma}, \tag{5.3.8}$$

one finds at $\sigma \to -\infty$:

$$\partial_\sigma^2 \varphi = \partial_\tau^2 \varphi , \tag{5.3.9}$$

$$\varphi = \varphi_{out}(\tau - \sigma) + \varphi_{in}(\tau + \sigma). \tag{5.3.10}$$

This means that the boundary $\sigma = -\infty$ is open! An infinite world of particles, on their way in or out, exists in the region $\zeta \cong 0$.

One would conclude from this that black holes are in a fundamental way different from other soliton like configurations in gauge theories such as magnetic monopoles: even if we would enclose the system in a finite space surrounding the black hole, particles near the black hole will form a continuous spectrum because they occupy a non-compact space.

It should be clear however that particles near $\sigma \to -\infty$ probe the infinitely small distance regime. Therefore, since quantum gravity is non-renormalizable, there will be values for σ beyond which the above analysis fails. Can gravitational forces turn the spectrum into a discrete one? The author believes this to be the case. But first we will explain what Hawking's result implies for this boundary.

5.4 The Hawking Effect

In this section we do not give a derivation, but rather a summary of the arguments that lead to the Hawking effect. Consider the region around the point $\mathcal{S}$ in fig. 5.1. Nothing in that region in space and time indicates that a black hole is about to form. Although there may be a lot of matter present in that region, it is still described by quite ordinary equations of state. In particular, one does not expect to see particles there moving at energies very far beyond $k_B T$, where T is the local temperature and k_B the Boltzmann constant.

But any particle that is very close to $\mathcal{S}$ and wants to escape to the external world needs such enormous energies. Information received from the black hole at late times *all* comes from a region extremely close to $\mathcal{S}$. Since we just decided that no particle could make it to the outside world, one would expect *no* information from the black hole: it should be truly black.

There are several ways to argue that in a quantum mechanical world things will be different. One way of saying is that if we concentrate on particles confined in a very small region, then the uncertainty relation will give these particles large fluctuations in their energies. Others argue that in the extreme gravitational field of the horizon pair creation will take place: positive energies will escape outwards, negative energies will live for a short moment and then be absorbed by the singularity at $r = 0$.

It is better to look at how a precise calculation goes [6]. One considers the complete set of solutions of the linearized field equations, and concentrates on those waves that have a significant amplitude at $\mathcal{S}$ and that do escape later from the horizon. The waves are then Fourier transformed with respect to that variable that locally plays the role of time. The positive frequency Fourier modes $a(\omega)$, $\omega > 0$, correspond to operators in Hilbert space that *annihilate* particles with the corresponding frequency; the negative frequency components $a(-\omega) = a^\dagger(|\omega|)$, are *creation* operators.

This holds both for the region near $\mathcal{S}$ and for the wave that later detaches from the horizon. However, the coordinate used as time is different in the two cases. Therefore, as the wave propagates positive and negative frequencies mix. The operator that annihilates a particle as seen near $\mathcal{S}$ will be a quantum mechanical mixture of an annihilation operator and a creation operator for the observer who later detects a particle as it leaves the black hole.

The *vacuum* is defined as the state $|\Omega\rangle$ that obeys

$$a(\omega)|\Omega\rangle = 0, \qquad \text{for all } \omega > 0. \tag{5.4.1}$$

If this were the description of the state near $\mathcal{S}$ then one finds after some calculations that the late observer will have

$$a_I(\omega)|\Omega\rangle \;+\; e^{-4\pi M\omega}\, a^\dagger_{II}(\omega)|\Omega\rangle = 0\,; \tag{5.4.2}$$

$$a_{II}(\omega)|\Omega\rangle \;+\; e^{-4\pi M\omega} a^\dagger_I(\omega)|\Omega\rangle \;= 0\,. \tag{5.4.3}$$

because he uses different annihilation operators a_I, a_{II}. What complicates matters is that those particles that are annihilated by a_{II} are invisible to that observer because they are the ones that are swallowed by the black hole. The solution of eqs. (5.4.2),(5.4.3) (for a given particle type at given angular momentum) is

$$|\Omega\rangle = \Pi_\omega \sqrt{1 - e^{-8\pi M\omega}} \sum_{n=0}^{\infty} e^{-4\pi M\omega n} |n\rangle_{I,\omega} |n\rangle_{II,\omega}\,. \tag{5.4.4}$$

Here and before, M is the black hole mass in natural units. The square root is a normalization factor.

We now say that the states in Hilbert space II act as a heat bath: they are not directly detected. Any observable operator only acts in I. As a consequence, expectation values are described by a density matrix:

$$\rho = \Pi_\omega \rho_\omega \qquad ; \qquad \rho_\omega(n, n') = (1 - e^{-8\pi M\omega})\, e^{-8\pi M\omega n}\, \delta_{n,n'} \; . \tag{5.4.5}$$

This corresponds exactly with a beam of radiated particles at a black body temperature given by

$$k_B T = \frac{1}{8\pi M} \; . \tag{5.4.6}$$

This is what Hawking's observation tells us about the boundary condition at the horizon: a thermal mixture of particles will come out. The physical equations at the boundary remain the same. The spectrum of states is still continuous.

5.5 The Black Hole Spectrum and Thermodynamics

The existence of an infinite spectrum of plane waves to and from the horizon, described by eq. (5.3.10), suggests that the spectrum of black holes is continuous. Yet, a more compelling reason exists to suggest that the spectrum should be discrete. A continuous spectrum would imply that an infinite number of mutually orthogonal states exists that can be stored within the volume defined by, say, the sphere $r = 4M$, with energies E between M and $M + \mathrm{d}M$, where M is the black hole mass. A small region of space would allow an infinite amount of information to be stored in there, at the cost of virtually no energy. This, we think, sounds unlikely.

Suppose for a moment that the density of black hole states at mass M (and charge Q and angular momentum L), is given by some finite number $\rho(M)$ (or $\rho(M, Q, L)$). Let us now compare the absorption process of an object with energy ΔE by a black hole with mass M, with the Hawking emission process for the same object by a hole with mass $M + \Delta E$:

$$(M) + (\Delta E) \Leftrightarrow (M + \Delta E) \; . \tag{5.5.1}$$

The absorption process has a cross section σ given approximately by

$$\sigma \simeq \pi R^2 \; , \tag{5.5.2}$$

where R is the black hole radius, $R = 2M$. The emission probability W is approximately

$$W \simeq \pi R^2 \, \rho_1(\Delta E)\, e^{-\beta_H \Delta E} \; , \tag{5.5.3}$$

where β_H is the inverse Hawking temperature, $\beta_H = 8\pi M$ (putting the gravitational constant κ and Boltzmann's constant k_B equal to one). $\rho_1(\Delta E)$ is the density of states for the objects radiated out with energy ΔE.

If there were a quantum mechanical theory for the black hole, the same quantities could be expressed in terms of transition amplitudes, using the "golden rule":

$$\sigma = |\langle M + \Delta E|T|M, \Delta E\rangle|^2 \rho(M + \Delta E) ; \tag{5.5.4}$$

where T is the transition matrix, and

$$W = |\langle M, \Delta E|T|M + \Delta E\rangle|^2 \rho(M)\rho_1(\Delta E) . \tag{5.5.5}$$

By virtue of PCT invariance, the matrix elements in (5.5.4) and (5.5.5) would be each other's conjugate, and therefore we find

$$\frac{\rho(M + \Delta E)}{\rho(M)} = \rho_1(\Delta E) \frac{\sigma}{W} = e^{\beta_H \Delta E} ; \tag{5.5.6}$$

this should hold for a range of values for ΔE as long as $\Delta E \ll M$, and with $\beta_H = 8\pi M$ we find

$$\rho(M) = C\ e^{4\pi M^2}, \tag{5.5.7}$$

where the universal constant C is the only unknown. Note that the exponent is one quarter of the area of the horizon, this is what one also finds in more general cases.

C could be finite, in which case we indeed have a finite spectrum density, or C is infinite, but in that case equations (5.5.4) and (5.5.5) could be considered for the lightest state in the continuum: $\rho(M + \Delta E) \stackrel{?}{=} \infty$; $\rho(M) \stackrel{?}{=}$ finite. Either this would be an absolutely stable thing ($W = 0$), in which case *virtual pair creation* of these objects would give infinite contributions to graviton self-energy diagrams, or the cross section σ would tend to infinity. Neither of these latter options sound physically very attractive, which is why we suspect C to be finite. We must realize however that very large numbers are indigenous in quantum gravity. It could be that C is of the order of 10^{40} or 10^{-40}.

It is unlikely that C is exactly constant. There may well be subdominant corrections either in the exponent, or in the form of powers in front of the exponent.

There is a different way to derive the same expression (5.5.7) for $\rho(M)$. This is by applying *thermodynamics* to the black hole [7]. The free energy F is defined by

$$e^{-\beta F} = \int_M^{M+dM} \rho(M)\mathrm{d}M\; e^{-\beta_H M} \simeq \rho(M)\; e^{-8\pi M^2}, \tag{5.5.8}$$

where the integral could be suppressed because the leading term is given by the exponent anyhow[3]. We inserted Hawking's value for the temperature and used natural units for the mass (the unit is the Planck mass, which is about 21.7 μg).

The expectation value for the energy M is

$$\langle M \rangle = \frac{\partial}{\partial\beta}(\beta F) = \frac{\beta}{8\pi}\,. \tag{5.5.9}$$

We conclude that

$$\beta F = \frac{\beta^2}{16\pi} = 4\pi M^2\,; \tag{5.5.10}$$

$$\rho\; = 4\pi M^2. \tag{5.5.11}$$

The philosophy used is that black holes behave just like little containers with some gas or liquid inside. If these communicate thermally with the outside world, one can deduce information concerning the total number of states from these reactions. The only thing slightly unusual about the thermal black holes is their negative specific heat:

$$c = \frac{\partial}{\partial T}\langle M \rangle = -\beta^2 \frac{\partial}{\partial \beta}(\frac{\beta}{8\pi}) = -\frac{\beta^2}{8\pi} = -8\pi M^2. \tag{5.5.12}$$

This implies that black holes cannot be completely in equilibrium with their environment. Martinez and York [8] speculate that this may have deep implications for the probabilistic interpretation of quantum mechanical expressions concerning black holes.

If indeed $\rho(M)$ is finite then this has also consequences for the symmetry aspects of a quantum gravity theory: we cannot have any absolute global conservation laws such as baryon number conservation! [9] The reason for this is that one could imagine dropping an unlimited amount of baryons into the hole, waiting each time for an equal amount of energy to reemerge in the form of Hawking radiation. Our derivation implies that equal numbers of baryons and anti-baryons should come out, on the average. So one could increase the hole's baryon number indefinitely. If ρ is finite there is however only a finite number of black hole states so that baryon number sooner or later becomes an ill-specified quantum number. Any preference of the hole to emit baryons rather than antibaryons would contradict the derivations mentioned in section 5.4.

[3]The necessity to provide for boundaries to the integral (5.5.8) stems from a fundamental thermal instability, see eq. (5.5.12).

Now this remark implies that applying quantum field theory alone to the black hole horizon will not yield us further details concerning these spectral states. We could have global symmetries in this quantum field theory, and they will be standing in the way! Something fundamental is missing.

Our preliminary standpoint will be the following: our present understanding of the laws of physics is imprecise: applying them gives us only statistical information about the black hole states. Since this statistical answer was a consequence of our applying quantum mechanics this may well mean that the usual statistical interpretation of quantum mechanics may be a consequence of a fundamental lack of precision. Imagine just a slight "uncertainty" in the Hamiltonian,

$$H \to H + \delta H, \tag{5.5.13}$$

where δH has some *probabilistic* distribution. In that case the solutions to the Schrödinger equation,

$$\frac{\mathrm{d}\psi}{\mathrm{d}t} = -iH\psi, \tag{5.5.14}$$

will have a "thermal" distribution, to be described by a density matrix [10], just as what one gets as Hawking radiation, eq. (5.4.5).

5.6 The Horizon Displacements

Rather than having mutually independent waves of ingoing and outgoing particles, we expect that in reality the outgoing radiation will be *determined* by what comes in via an S-matrix. The first thing to suspect is that it was wrong to neglect gravitational interactions between ingoing and outgoing radiation. What will these interactions be like? Fortunately, they can be computed rather precisely.

Consider a very light particle falling in the black hole at time $t = t_1$. Its gravitational field is weak and can easily be obtained by linearizing Einstein's equations. Now consider an observer at a much later time $t = t_2$. In his coordinate frame he will put the origin at t_1. If now he makes the transition to Kruskal coordinates to describe the metric surrounding him, he will have typically (see eq. (5.2.2))

$$\frac{x}{y} = -e^{(t-t_2)/2M}. \tag{5.6.1}$$

Thus, his x coordinate will shrink and his y coordinate will blow up. Even the slightest disturbance our light proton gave to movements of other particles in the y direction will sooner or later become important. In particular, the position of the horizon with respect to the x axis will change. The shift δy of the horizon will be a function of the angular coordinate $\Omega = (\vartheta, \varphi)$:

$$\delta y(\Omega) = f(\Omega, \Omega_1)\, p_{in} \,, \tag{5.6.2}$$

and a computation shows that the Green function f is a solution of

$$(1 - \Delta_\Omega)\delta y(\Omega) = 4\pi\kappa p_{in}\, \delta^2(\Omega, \Omega_1) \,, \tag{5.6.3}$$

where Δ_Ω is the angular Laplacian and Ω_1 is the set of angles at which the particle dropped in. p_{in} is the momentum of the infalling proton in the Kruskal frame. δy is uniquely determined by (5.6.3) because $1 - \Delta_\Omega$ has a unique inverse.

Eq. (5.6.3) is found by writing the space-time metric obtained from gluing the two Schwarzschild solutions together as an *Ansatz* and then imposing Einstein's equations [11]. The Green function f can be given in an integral form but this is not very illuminating. One finds that $f(\Omega, \Omega_1) > 0$ for all Ω, Ω_1, and f diverges logarithmically when $\Omega \to \Omega'$.

A consequence of these observations is that if we drop a particle into the black hole, the position of the horizon at times t *before* the particle fell in, changes, as drawn in fig. 5.3. This change is barely perceptible at times $t \lesssim t_1$, but at times $t_2 \gg t_1$ the change is large. An observer there sees Hawking radiation that now originated in a *different* region of space-time than it would have if the particle had not been thrown in.

Is this consequence of any importance? What does it matter if the Hawking radiation originated somewhere else? It will certainly look the same as before.

We will argue in the next section that only in a quantum theory that is detailed enough to give us a scattering matrix instead of a density matrix, shifting horizons will be relevant. Indeed, important constraints on the scattering matrix will be found.

5.7 The Scattering Matrix. Black Hole – White Hole Duality

Suppose a scattering matrix exists. This means if we have completely specified the state $\{p_1, p_2, \ldots\}$ of all particles that ever went into the black hole, the outgoing matter should be in one well-specified state $|\psi\rangle_{out}$. A basis for $|\psi\rangle_{out}$ is the set of states where all outgoing particles have well-specified momenta at a certain time $t = t_0$. Now at $t = t_1 > t_0$ we drop a light particle into the hole, with momentum p_{in} (in regular coordinates) at solid angle Ω_1. The change this induces for the outgoing wave is now determined primarily by the horizon

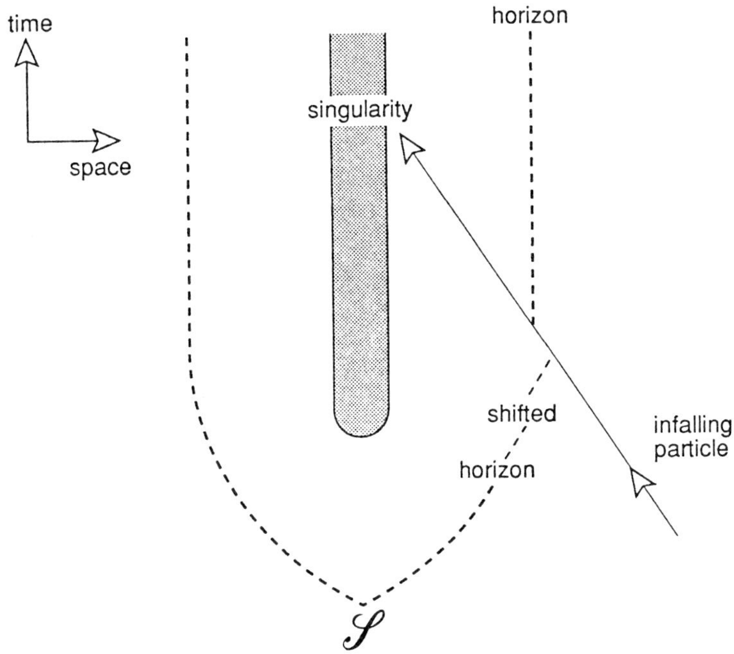

Figure 5.3: *The shifting horizon.*

shift (if other, non-gravitational interactions may be ignored). Thus, the new state will now be

$$|\psi\rangle_{out} \rightarrow e^{-i\int P_{out}(\Omega)\delta y(\Omega)\,\mathrm{d}^3\Omega}\,|\psi\rangle_{out}\ , \tag{5.7.1}$$

where $P_{out}(\Omega)$ is the operator that generates a shift in the configurations at the solid angle Ω. It is, of course, also the total momentum emerging at solid angle Ω. In here, we can now substitute eq. (5.6.2) for δy.

Now this means that if we know $|\psi\rangle_{out}$ at one stage, then $|\psi\rangle_{out}$ can, in principle, be determined after allowing any number of particles to fall in. If we may ignore non-gravitational interactions, we see that all states $|\psi\rangle_{out}$ ever to be produced by the black hole are generated by the operator $P_{out}(\Omega)$ from one single state. Therefore, $|\psi\rangle_{out}$ must be generated by the *algebra* of these operators. Similarly, the *ingoing* particles are only distinguished by the *total* momentum $p_1(\Omega_1)$ at each solid angle Ω_1.

We find the following important result [12]. For the incoming wave functions one may diagonalize the operators $P_{in}(\Omega_1) = p_{in}(\Omega_1)$, and for the outgoing states we diagonalize $P_{out}(\Omega)$. Eq. (5.7.1) then tells us how a change in p_{in} affects the outgoing state. Up to a proportionality factor, the complete transformation rule for ingoing states into outgoing states should be generated by this equation. This rule is not difficult to find:

$$\langle\{p_{out}(\Omega)\}|\{p_{in}(\Omega')\}\rangle = N\, e^{-i\int d^2\Omega\, d^2\Omega'\, p_{out}(\Omega)\, f(\Omega,\Omega')\, p_{in}(\Omega')}, \tag{5.7.2}$$

where N is a normalization factor.

Eq. (5.7.2) is the S-matrix we wanted. *If* an S-matrix exists, and if we may ignore other than the longitudinal gravitational forces, it must be this one.

The problem with it is the algebra that generated the basis in which it is defined. We have the following commutation rules

$$[p_{in}(\Omega), p_{in}(\Omega')] = 0\ ; \qquad [p_{in}(\Omega), x_{in}(\Omega')] = -i\delta^2(\Omega,\Omega')\ ; \tag{5.7.3}$$

$$[p_{out}(\Omega), p_{out}(\Omega')] = 0\ ; \qquad [p_{out}(\Omega), y_{out}(\Omega')] = -i\delta^2(\Omega,\Omega') \tag{5.7.4}$$

and we have the relation

$$y_{out}(\Omega) = \int d^2\Omega_1\; f(\Omega,\Omega_1)\; p_{in}(\Omega_1). \tag{5.7.5}$$

This implies

$$[x_{in}(\Omega), y_{out}(\Omega')] = i f(\Omega,\Omega'), \tag{5.7.6}$$

so that we have also

$$x_{in}(\Omega) = -\int d^2\Omega'\; f(\Omega,\Omega')\; p_{out}(\Omega'). \tag{5.7.7}$$

The operators x_{in} and y_{out} could be interpreted as coordinates of "particles", but then there should be *exactly one particle at every value of* Ω. This is where this Hilbert space differs from ordinary Fock space, where we may have any number of particles (mostly this will be zero) at every mode. This is also why it will be difficult to interpret our S-matrix directly as a matrix describing scattering of familiar particles.

More appropriately, x_{in} and y_{out} are to be seen as the position operators for the *past* and the *future* horizon. We then recognise an important consequence of our description of Hilbert space: past and future horizons cannot both be localized accurately; these obey an uncertainty relation. Indeed, they are each other's dual conjugates, much in the same way as coordinates are dual to momenta.

We claim that this also does away with a question considered often in the literature: does the time-reversed black hole (called "white hole") exist? Does the "eternal black hole" (one with a past white hole that was already there before the universe began) exist? The questions are not appropriate if our S-matrix exists: white hole coordinates are ill-specified once a black hole was localised.

The difficult but perhaps exciting picture that emerges is that the exact shape of either the past or the future horizon may completely determine the particle content of the black hole's vicinity. It should be possible to refine this picture by incorporating gravitational forces in the transverse direction, and non-gravitational forces. It is not hard to take the electromagnetic force into account. Here, the electric charge density operator $\rho(\Omega)$ and the gauge phase operator $\phi(\Omega)$ are each other's duals. Electromagnetic shock waves (Čerenkov radiation) surrounding charged massless particles are very similar to gravitational shock waves.

As will be explained in the next section, we expect a cut-off in Ω space. Probably the transverse forces are responsible for that. Surely, if coordinates in the transverse direction would be specified with accuracies better than the Planck length (implying $\delta\Omega \lesssim M_{pl}/M$, where M is the black hole mass), then momenta in the transverse direction exceed the Planck energy so that also shifts in the transverse direction will arise that are bigger than $\delta\Omega$. The simplest cut-off would be a lattice in Ω space, but reality will probably be more complicated.

5.8 Discrete Physics

Imagine a volume $V = L^3$ in three-dimensional space. Suppose that one could define a quantum (field) theory of all phenomena (particles, black holes) inside this volume, in terms of a Hilbert space. We ask: how many dimensions (basis elements) $\mathcal{D}$ does this Hilbert space have as V grows? In ordinary field theories the answer is strictly infinite. Even a single particle can occupy infinitely many states, so surely a second-quantized theory will have an infinite-dimensional Hilbert space. Only if we would introduce a rigid lattice cut-off, and accept only fundamental fermions in our theory, the total dimensionality would be finite. It would grow like

$$\mathcal{D} \cong \mathcal{O}(e^{\wedge^3 L^3}), \tag{5.8.1}$$

where $\wedge$ is the inverse lattice size.

But this changes if the gravitational force is taken into account. In that case we cannot allow the total energy to exceed a certain value depending on V:

$$E_{tot} \leq \frac{L}{2\kappa}, \tag{5.8.2}$$

because otherwise a black hole would form with a horizon that stretches beyond the edges of V.

Most quantum field theories have only

$$\mathcal{D} \cong \mathcal{O}(e^{C\kappa^{-3/4}L^{3/2}}) \tag{5.8.3}$$

basis elements with energy less than that (C is some constant). If we allow black holes we expect more states. Suppose there are N black holes with labels $i = 1, \ldots, N$. Each black hole has a density of states of the order of

$$\mathcal{D}_i \cong e^{4\pi M_i^2}, \tag{5.8.4}$$

(leaving again ${M_{pl}}^2 = \kappa = 1$), and the total system has

$$\mathcal{D} = \Pi_i \mathcal{D}_i = e^{4\pi \Sigma_i {M_i}^2} \leq e^{4\pi(\Sigma_i M_i)^2}; \tag{5.8.5}$$

whereas the total energy is

$$E_{tot} = \sum_i M_i \leq \tfrac{1}{2} L. \tag{5.8.6}$$

Therefore, allowing black holes we find

$$\mathcal{D} \lesssim e^{L^2 \pi}, \tag{5.8.7}$$

which, surprisingly, grows exponentially with the surface area L^2 rather than the volume L^3.

We claim that in any "complete" quantum theory of the world the total dimensionality of Hilbert space inside a volume L^3 must be finite and approximately given by eq. (5.8.7). Theories where this dimensionality does not grow exponentially with the volume but with the surface area are not easy to construct. It could mean that there is some sort of additional constraint to be imposed on all states in the "physical part" of Hilbert space, whose solutions would have the dimensionality given by (5.8.7).

There is an interesting class of finite-dimensional theories. These are the so-called cellular automata. A cellular automaton is a system containing a definite number of completely discrete and limited variables located at "cells". The contents of each cell are continuously updated according to a given arithmetic rule depending on the previous values and the values inside neighbouring cells. They are ideal for computer simulations.

A cellular automaton may either be completely deterministic or quantum mechanical. In a quantum mechanical cellular automaton the updating is prescribed by some unitary operator U defined in Hilbert space. At each tick of the clock we have

$$|\psi(t+1)\rangle = U|\psi(t)\rangle. \tag{5.8.8}$$

This equation is the direct discrete analogon of the Schrödinger equation. The model is *deterministic* if a basis exists in terms of which the operator U happens to be also a permutation operator:

$$U|n_1, n_2, \ldots\rangle = |P(n_1, n_2, \ldots)\rangle. \tag{5.8.9}$$

If ever we find the operator U describing the real world it might not be easy to establish whether or not it can be seen to be a deterministic one, because the search for the corresponding basis may be difficult. It is very tempting to *suspect* that U should be deterministic. It then remains to be seen how we can understand that nevertheless features typical for quantum mechanics govern the macroscopic world. Experiences with computer simulations with cellular automata show that they often behave *chaotically*. This means that there is no simple way to describe the macroscopic world in terms of deterministic laws in spite of local determinism. Perhaps the *only* way to describe the macroscopic world is *via* the Schrödinger equation (5.8.8) or in other words: perhaps our world is a cellular automaton but it allows only a statistical description. The mathematics of this statistics may happen to be that of conventional quantum mechanics [13].

There are numerous difficulties with such a picture but we think it cannot be ruled out. It would be a neat way to resolve the philosophical difficulties usually caused by quantum mechanics. In particular we will have to face the Einstein Podolski Rosen paradox [14] and the question how one can understand the violation of Bell's inequalities [15] by quantum mechanical interference effects. We stress however that strictly speaking there is no logical contradiction here at all because the *vacuum state* $|\phi\rangle$ will have to be a superposition of all basis elements, in particular if we use the deterministic evolution operator P of eq. (5.8.9). Since all our experimental set-ups are always surrounded by a vacuum we should not be surprised to find "interference effects".

In attempting to improve our calculation of the S-matrix elements (5.7.2) we found that the horizon can probably not be kept topologically as simple as an S_2 sphere. Indeed at the point $\mathcal{S}$ in fig. 5.1 the topology must be very involved. One might speculate that the topological structures at $\mathcal{S}$ are denumerable and this could be a starting point for a completely discrete theory for the black hole horizon, and, with that, for space-time itself.

Acknowledgement

The author thanks Duke University, Durham NC, for its hospitality.

References

[1] K.S. Thorne, *Black Holes: The Membrane Paradigm*, Yale Univ. Press, New Haven, 1986; S. Chandrasekhar, *The Mathematical Theory of Black Holes*, Clarendon Press, Oxford University Press.

[2] S.W. Hawking, Comm. Math. Phys. **43** (1975) 199; J.B. Hartle and S.W. Hawking, Phys. Rev. **D13** (1976) 2188; W.G. Unruh, Phys. Rev. **D14** (1976) 870; S.W. Hawking and G. Gibbons, Phys. Rev. **D15** (1977) 2738; S.W. Hawking, Comm. Math. Phys. **87** (1982) 395.

[3] G. 't Hooft, J. Geom. and Phys. **1** (1984) 45.

[4] S.W. Hawking and G.F.R. Ellis, *The Large-scale Structure of Space-time*, Cambridge Univ. Press, 1973; C.W. Misner, K.S. Thorne and J.A. Wheeler, *Gravitation*, Freeman, San Francisco, 1973.

[5] W. Rindler, Am. J. Phys. **34** (1966) 1174.

[6] See ref. [2] and also G. 't Hooft, Acta Phys. Polon. **B19** (1988) 187.

[7] J.D. Bekenstein, Nuovo Cim. Lett. **4** (1972) 737; Phys. Rev. **D7** (1973) 2333; Phys. Rev. **D9** (1974) 3292.

[8] E.A. Martinez and J.W. York, Jr., Chapel Hill preprint (1989).

[9] J.D. Bekenstein, Phys. Rev. **D5** (1972) 1239, 2403.

[10] S.W. Hawking, Comm. Math. Phys. **87** (1982) 395; S.W. Hawking and R. Laflamme, Phys. Lett. **B209** (1988) 39.

[11] T. Dray and G. 't Hooft, Nucl. Phys. **B253** (1985) 173.

[12] G. 't Hooft, in: *Proceedings of the 4th seminar on Quantum Gravity*, May 25–29, 1987, Moscow, USSR, ed. by M.A. Markov et al., World Scientific, 1988, pp. 551–567.

[13] G. 't Hooft, J. Stat. Phys. **53** (1988) 323.

[14] A. Einstein, B. Podolski and N. Rosen, Phys. Rev. **47** (1935) 777; M. Jammer, *The Conceptual Developments of Quantum Mechanics*, Mc. Graw-Hill, 1966.

[15] J.S. Bell, Physics **1** (1964) 195.

Between Science and Technology
A. Sarlemijn and P. Kroes (Editors)

Chapter 6

La nouvelle cuisine

J.S. Bell

CERN, Geneva

Respectfully dedicated to the great chef

6.1 Introduction

There is an ongoing series of symposia, at Tokyo, on "Foundations of Quantum Mechanics in the Light of New Technology" [1,2]. Indeed new technology (electronics, computers, lasers,...) has made possible new demonstrations of quantum queerness. And it has made possible practical approximations to old gedankenexperiments. Over the last decade or so have appeared beautiful experiments [1,2] on "particle" interference and diffraction, with neutrons and electrons, on "delayed choice", on the Ehrenburg–Siday–Aharonov–Bohm effect, and on the Einstein–Podolsky- Rosen–Bohm correlations. These last are of particular relevance for the particular themes of this paper. But those themes arise already in the context of technology which is neither new nor advanced, as is illustrated by the following passage:

> "I want to boil an egg. I put the egg into boiling water and I set an alarm for five minutes. Five minutes later the alarm rings and the egg is done. Now the alarm clock has been running according to the laws of classical mechanics uninfluenced by what happened to the egg. And the egg is coagulating according to laws of physical

chemistry and is uninfluenced by the running of the clock. Yet the coincidence of these two unrelated causal happenings is meaningful, because I, the great chef, imposed a structure on my kitchen.", H.B.G. Casimir [3].

These notions, of cause and effect on the one hand, and of correlation on the other, and the problem of formulating them sharply in contemporary physical theory, will be the themes of my talk. I will be particularly concerned with the idea that effects are near to their causes:

> "If the results of experiments on free fall here in Amsterdam would depend appreciably on the temperature of Mont Blanc, on the height of the Seine below Paris and on the position of the planets, one would not get very far.", H.B.G. Casimir [4].

Now at some very high level of accuracy, all these things *would* become relevant for free fall in Amsterdam. However even then we would expect their influence to be retarded by at least the time that would be required for the propagation of light. I will be much concerned here with the idea of the velocity of light as a limit. What exactly does it limit?

6.2 What can not go faster than light?

Once when I arrived to give a talk on this topic, at the University of Hamburg, it was pointed out to me that some graffiti had been added to one of the announcements (figure 6.1). To the question "What can not go faster than light ?" had been volunteered the reply: "John Bell, for example". I have been wondering ever since what exactly that meant. When I walk, one foot remains planted on the ground, while the other advances. When I talk, I wave my hands (as you will see) and they have different velocities — from one another and from my head. Perhaps what was meant was that no *part* of John Bell can go faster than light. But that raises the question of how far I can be resolved into parts ... legs and arms, fingers and toes ... cells ... molecules, atoms ... electrons. Was it meant that none of my *electrons*, for example, go faster than light?

The idea that no particle could go faster than light arose late in the nineteenth century, first for electrically charged particles, and then for all particles. Even then the "particles" were envisaged as extended, and questions about their "parts" could be posed ... And now the sharp localization of objects in classical theory has been replaced by the fuzziness of wave mechanics and the complications of quantum field theory. The concept "velocity of an electron" is now unproblematic only when not thought about.

Donnerstag, 21. 1. 1988
16.45 Uhr
Hörsaal I
Institut für Angewandte Physik
Jungiusstr. 11

J. Bell (CERN)

What cannot go faster than light ?

John Bell, for example!

Figure 6.1: *"What can not go faster than light?"*

The situation is further complicated by the fact that there *are* things which *do* go faster than light. British sovereignty is the classical example. When the Queen dies in London (may it long be delayed) the Prince of Wales, lecturing on modern architecture in Australia, becomes *instantaneously* King. (Greenwich Mean Time rules here). And there are things like that in physics. In Maxwell's theory, the electric and magnetic fields in free space satisfy the wave equation:

$$\frac{1}{c^2}\frac{\partial^2 \mathbf{E}}{\partial t^2} - \nabla^2 \mathbf{E} = 0 \,, \tag{6.2.1}$$

$$\frac{1}{c^2}\frac{\partial^2 \mathbf{B}}{\partial t^2} - \nabla^2 \mathbf{B} = 0 \tag{6.2.2}$$

...corresponding to propagation with velocity c. But the scalar potential, if one chooses to work in "Coulomb gauge", satisfies Laplace's equation

$$-\nabla^2 \phi = 0 \tag{6.2.3}$$

...corresponding to propagation with infinite velocity. Because the potentials are only mathematical conveniences, and arbitrary to a high degree, made definite only by the imposition of one convention or another, this infinitely fast propagation of the Coulomb-gauge scalar potential disturbs no one. Conventions can propagate as fast as may be convenient. But then we must distinguish in our theory between what is convention and what is not.

6.3 Local beables

No one is obliged to consider the question "What can not go faster than light?". But if you decide to do so, then the above remarks suggest the following: you must identify in your theory "local *be*ables". The *be*ables of the theory are those entities in it which are, at least tentatively, to be taken seriously, as corresponding to something real. The concept of "reality" is now an embarrassing one for many physicists, since the advent of quantum mechanics, and especially of "complementarity". But if you are unable to give some special status to things like electric and magnetic fields (in classical electromagnetism), as compared with the vector and scalar potentials, and British sovereignty, then we cannot begin a serious discussion. *Local* beables are those which are definitely associated with particular space-time regions. The electric and magnetic fields of classical electromagnetism, $\mathbf{E}(t,x)$ and $\mathbf{B}(t,x)$, are again examples, and so are integrals of them over limited space-time regions. The total energy in all space, on the other hand, may be a beable, but is certainly not a local one.

Now it may well be that there just *are* no local beables in the most serious theories. When space-time itself is "quantized", as is generally held to be necessary, the concept of locality becomes very obscure. And so it does also in presently fashionable "string theories" of "everything". So all our considerations are restricted to that level of approximation to serious theories in which space-time can be regarded as given, and localization becomes meaningful. Even then, we are frustrated by the vagueness of contemporary quantum mechanics. You will hunt in vain in the text-books for the local *be*ables of the theory. What you may find there are the so-called "local observables". It is then implicit that the apparatus of "observation", or, better, of experimentation, and the experimental results, are real and localized. We will have to do as best we can with these rather ill-defined local beables, while hoping always for a more serious reformulation of quantum mechanics where the local beables are explicit and mathematical rather than implicit and vague.

6.4 No signals faster than light

The concept of particle is no longer sharp, so the concept of particle velocity is not sharp either. The answer to our question can no longer be: "particles can not go faster than light". But perhaps it could still be: "cause and effect". As far as I know, this was first argued by Einstein, in the context of special relativity theory. In 1907 he pointed out [5] that if an effect followed its cause sooner than light could propagate from the one place to the other, then in some other inertial frames of reference the "effect" would come before the "cause"! He wrote [6]

> "...in my opinion, regarded as pure logic ... it contains no contradictions; however it absolutely clashes with the character of our total experience, and in this way is proved the impossibility of the hypothesis ..."

of a causal chain going faster than light.

The kind of thing that Einstein found unacceptable is illustrated in figure 6.2. If I had a "tachyon" gun, i.e. one that could shoot bullets (or rays, or whatever) faster than light, then I could commit a murder without fear of punishment. This could be done by exploiting the relativity of time. I would lure my victim to the origin of coordinates O. Then I would run rapidly past, pulling the trigger at the appropriate moment P, shortly before time $t' = 0$ on my watch, and the deed would soon be done at time $t' = 0$. This would also be (by hypothesis) time $t = 0$, where t is Greenwich Mean Time, as used (at least during the winter in England) by the police, the courts of justice, and indeed all other institutions firmly planted on the English ground. But at time $t = \epsilon$ (where ϵ as usual is very small) the trigger has not yet been pulled, although the victim is dead. Indeed from this earthly point of view what happens at the origin of coordinates is that the unfortunate victim collapses spontaneously, with the spontaneous emission of an antitachyon. Happening to be passing, I catch the antitachyon into the barrel of my gun, and so prevent possible injury to other passers-by. I should get a medal.

Even Einstein would have hesitated to accept such relativity of morality. Most citizens will feel that such actions, if not against the laws of the land, should be excluded by the laws of nature. What we have to do then is to add to the laws of relativity some responsible causal structure. To avoid causal chains going backward in time in some frames of reference, we require them to go slower than light in any frame of reference.

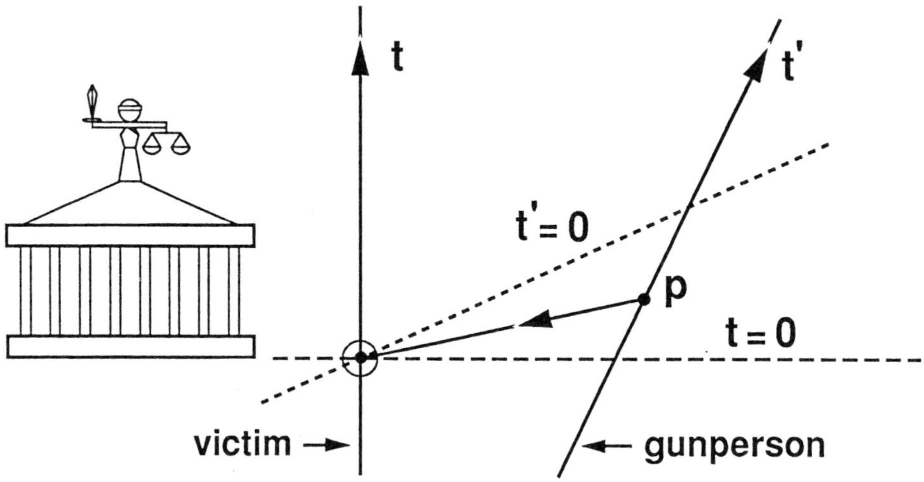

Figure 6.2: *Perfect tachyon crime.*

6.5 Local commutativity

Ordinary "local" quantum field theory does have a causal structure. As everyone knows, it gives rise to "dispersion relations". In their pioneering paper on dispersion relations in relativistic quantum field theory, Gell-Mann, Goldberger, and Thirring [7] write:

> "The quantum mechanical formulation of the demand that waves do not propagate faster than light is, as is well known, the condition that the measurement of two observable quantities should not interfere if the points of measurement are space-like to each other ... the commutators of two Heisenberg operators ... shall vanish if the operators are taken at space-like points."

Thus for Heisenberg operators A and B for space-time points x and y,

$$[A(x), B(y)] = 0, \text{ for } (x_0 - y_0)^2 < (\mathbf{x} - \mathbf{y})^2 \tag{6.5.1}$$

... which is called "local commutativity".

The only way that I know to relate local commutativity to any sort of causality concerns the response of the quantum system to external interventions. Two sorts of external intervention are contemplated in ordinary quantum mechanics. They are the making of "measurements", and the imposition of "external fields".

The "non-interference" of "measurements" of commuting "observables" includes the following: the probability of any particular result for one of them is unaltered by whether or not the other is indeed measured, when all possible results for the latter (if indeed measured) are averaged over [8]. And so, in a theory with local commutativity, an experimental physicist can not increase the probability that a rival will be "measured" as dead in a spacelike-separated region, — by himself or herself making "measurements". The last sentence illustrates, by the way, the grotesque misuse of the word "measurement" in contemporary quantum mechanics. The more careful writers use sometimes the word "preparation" instead, and this would be less inappropriate here for whatever action the gunperson might take towards the desired end. Those actions will be in vain, in a locally commutative theory, if like "measurements" and "preparations" they result only in the "collapse of the wavefunction" to an eigenstate of a nearby "observable".

An "external field" is a c-number field on which the theory imposes no restrictions, i.e. about which it asserts no laws. The Lagrangian can be allowed to depend on such fields. The arbitrariness of such fields can be supposed to represent the freedom of experimenters, for example to do one variation of an experiment rather than another. Consider the effect of a small variation of such a field ϕ. The variation of the Lagrangian density will be of the form

$$\delta L(y) = Y(y)\ \delta\phi(y)\ , \tag{6.5.2}$$

where, in "local" theory, $Y(y)$ is some operator belonging to the space-time point y. Then it is an easy exercise in quantum mechanics to show that for a Heisenberg operator $X(x)$, the retarded change is given by

$$\frac{\delta X(x)}{\delta\phi(y)} = i\theta(x_0 - y_0)\ [X(x), Y(y)]\ , \tag{6.5.3}$$

where θ is the step function, zero for negative argument. Then with local commutativity the statistical predictions of quantum mechanics, for "measurement results", do not depend on external fields outside the backward lightcone of the "observables" in question. So, no superluminal signalling with external fields.

6.6 Who could ask for anything more?

Could the no-superluminal-signalling of "local" quantum field theory be regarded as an adequate formulation of the fundamental causal structure of physical theory? I do not think so. For although "local commutativity" has a nice sharp-looking mathematical appearance, the concepts involved in relating it to causal structure are not very satisfactory.

This is notoriously so as regards the notion of "measurement" and the resulting "collapse of the wavefunction". Does this happen sometimes outside laboratories? Or only in some authorized "measuring apparatus"? And whereabouts in that apparatus? In the Einstein–Podolsky–Rosen–Bohm experiment, does "measurement" occur already in the polarizers, or only in the counters? Or does it occur still later, in the computer collecting the data, or only in the eye, or even perhaps only in the brain, or at the brain-mind interface of the experimenter?

The notion of external field is a more honourable one than that of "measurement". There are many cases in practice where an electromagnetic field can be considered, in an adequate approximation, to be classical and external to the quantum system. For example, a variation on the EPRB experiment involves neutral spin-half particles instead of photons. The polarization analyzers can then be Stern–Gerlach magnets, and their magnetic fields can be treated as "external" ... in a good approximation. But an accurate treatment of the electromagnetic field involves its incorporation into the quantum system. And must we not also so incorporate

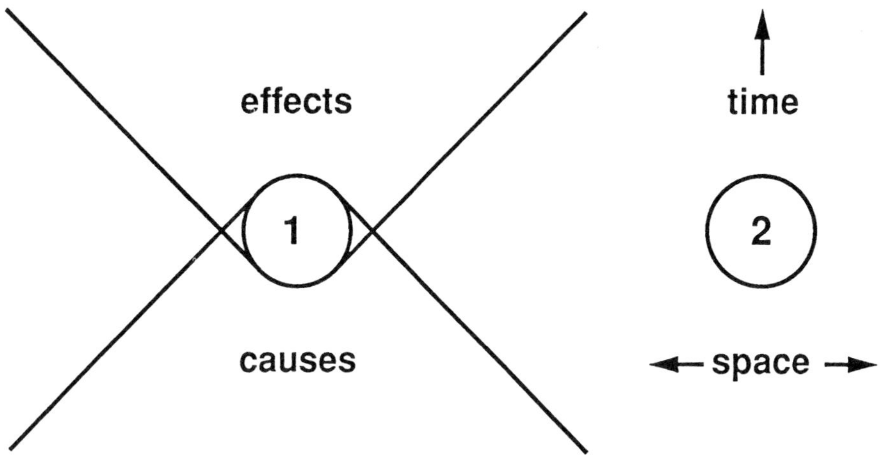

Figure 6.3: *Space-time location of causes and effects of events in region 1.*

the magnets, the hand of the experimenter, the brain of the experimenter? Where are truly "external" fields to be found? Perhaps at the interface between the brain and the mind?

Who am I to deny that a sharp formulation of causal structure in physical theory requires reference to the minds of experimental physicists? Or that there just *was* no causal structure before the emergence of that profession (this might have interesting implications in cosmology). But before trying to figure out from which parts of their heads, and when, the fundamental causal cones emerge, should we not look for alternatives?

As a first attempt let us formulate the following ...

6.7 Principle of local causality

> The direct causes (and effects) of events are near by, and even the indirect causes (and effects) are no further away than permitted by the velocity of light.

Thus for events in a space-time region 1 (figure 6.3) we would look for causes in the backward light cone, and for effects in the future light-cone. In a region like 2, space-like separated from 1, we would seek neither causes nor effects of events in 1. Of course this does not mean that events in 1 and 2 might not be correlated, as are the ringing of Professor Casimir's alarm and the readiness of his egg. They are two separate results of his previous actions.

The above principle of local causality is not yet sufficiently sharp and clean for mathematics.

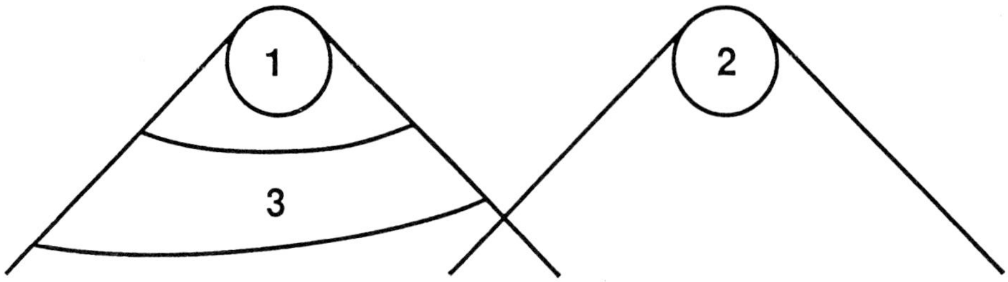

Figure 6.4: *Full specification of what happens in 3 makes events in 2 irrelevant for predictions about 1 in a locally causal theory.*

Now it is precisely in cleaning up intuitive ideas for mathematics that one is likely to throw out the baby with the bathwater. So the next step should be viewed with the utmost suspicion:

> A theory will be said to be locally causal if the probabilities attached to values of local beables in a space-time region 1 are unaltered by specification of values of local beables in a space-like separated region 2, when what happens in the backward light cone of 1 is already sufficiently specified, for example by a full specification of all local beables in a space-time region 3 (figure 6.4).

It is important that region 3 completely shields off from 1 the overlap of the backward light cones of 1 and 2. And it is important that events in 3 be specified completely. Otherwise the traces in region 2 of causes of events in 1 could well supplement whatever else was being used for calculating probabilities about 1. The hypothesis is that any such information about 2 becomes redundant when 3 is specified completely. The ringing of the alarm establishes the readiness of the egg. But if it is already given that the egg was nearly boiled a second before, then the ringing of the alarm makes the readiness no more certain.

Consider for example Maxwell's equations, in the source-free case for simplicity. The fields **E** and **B** in region 1 are completely determined by the fields in region 3, regardless of those in 2. Thus this is a locally causal theory in the present sense. The deterministic case is a limit of the probabilistic case, the probabilities becoming delta functions.

Note, by the way, that our definition of locally causal theories, although motivated by talk of "cause" and "effect", does not in the end explicitly involve these rather vague notions.

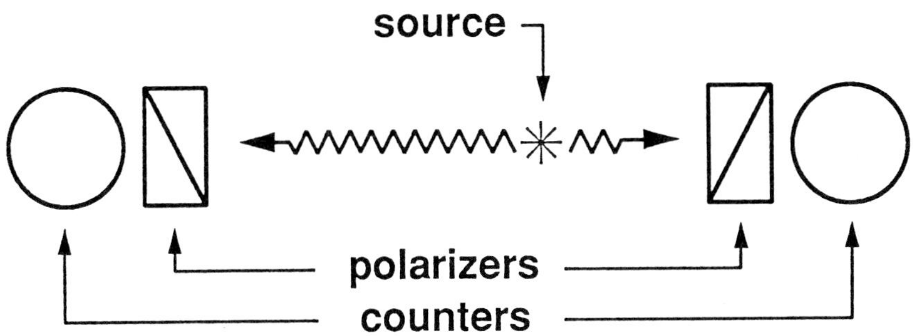

Figure 6.5: *Einstein–Podolsky–Rosen–Bohm gedankenexperiment.*

6.8 Ordinary quantum mechanics is not locally causal

That ordinary quantum mechanics is not locally causal was pointed out by Einstein, Podolsky and Rosen, in 1935 [9]. Their argument was simplified by Bohm [9] in 1951. Let the "source" in figure 6.5 emit a pair of photons in opposite directions along the z-axis. Let them be in joint polarization state

$$\frac{1}{\sqrt{2}}\{X(1)X(2)+Y(1)Y(2)\}\ , \tag{6.8.1}$$

where X and Y are states of linear polarization in x and y directions. Let the polarizers be so oriented as to pass the X states and block the Y's. Each of the counters considered separately has on each repetition of the experiment a 50% chance of saying "yes". But when one counter says "yes" so also always does the other, and when one counter says "no" the other also says "no", according to quantum mechanics. The theory requires a perfect correlation of "yeses" or "nos" on the two sides. So specification of the result on one side permits a 100% confident prediction of the previously totally uncertain result on the other side. Now in ordinary quantum mechanics there just *is* nothing but the wavefunction for calculating probabilities. There is then no question of making the result on one side redundant on the other by more fully specifying events in some space-time region 3. We have a violation of local causality.

Most physicists were (and are) rather unimpressed by this. That is because most physicists do not really accept, deep down, that the wavefunction is the whole story. They tend to think that the analogy of the glove left at home is a good one. If I find that I have brought only one glove, and that it is right-handed, then I predict confidently that the one still at home will be seen to be left handed. But suppose we had been told, on good authority, that gloves

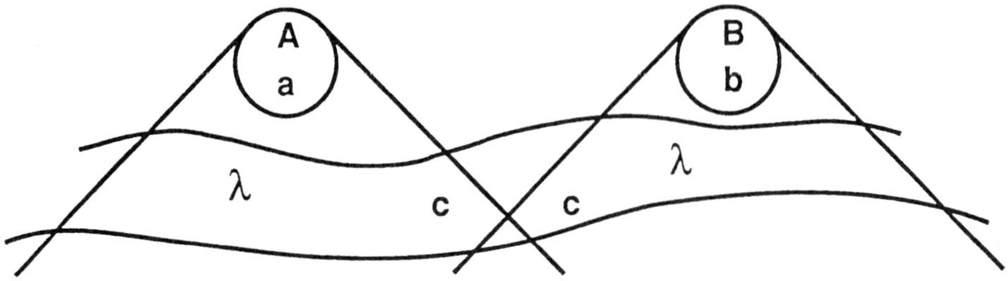

Figure 6.6: *Diagram for CHHS inequality derivation.*

are neither right- or left-handed when not looked at. Then that, by looking at one, we could predetermine the result of looking at the other, at some remote place, would be remarkable. Finding that this is so in practice, we would very soon invent the idea that gloves are already one thing or the other even when not looked at. And we would begin to doubt the authorities that had assured us otherwise. That common-sense position was that taken by Einstein, Podolsky and Rosen, in respect of correlations in quantum mechanics. They decided that the wavefunction, making no distinction whatever between one possibility and another, could not be the whole story. And they conjectured that a more complete story would be locally causal.

However it has turned out that quantum mechanics can not be "completed" into a locally causal theory, at least as long as one allows, as Einstein, Podolsky and Rosen did, freely operating experimenters. The analogy of the gloves is not a good one. Common sense does not work here.

6.9 Locally explicable correlations

In the space-time diagram of figure 6.6 we denote by A (=+1 or -1) the output from the counter on the left ("yes" or "no"). And B (=+1 or -1) is the output from the counter on the right. We denote by a and b the angles by which the polarizers are rotated from some standard positions in which they are parallel. We consider a slice of space-time 3 earlier than the regions 1 and 2 and crossing both their backward lightcones where they no longer overlap. In region 3 let c stand for the values of any number of other variables describing the experimental set-up, as admitted by ordinary quantum mechanics. And let λ denote any number of hypothetical additional complementary variables needed to complete quantum mechanics in the way envisaged by EPR. Suppose that the c and λ together give a complete specification of at least those parts of 3 blocking the two backward lightcones.

Let

$$\{A, B|a, b, c, \lambda\} \tag{6.9.1}$$

denote the probability of particular values A and B given values of the variables listed on the right. By a standard rule, the joint probability can be expressed in terms of conditional probabilities:

$$\{A, B|a, b, c, \lambda\} = \{A|B, a, b, c, \lambda\}\ \{B|a, b, c, \lambda\}\,. \tag{6.9.2}$$

Invoking local causality, and the assumed completeness of c and λ in the relevant parts of region 3, we declare redundant certain of the conditional variables in the last expression, because they are at space-like separation from the result in question. Then we have

$$\{A, B|a, b, c, \lambda\} = \{A|a, c, \lambda\}\ \{B|b, c, \lambda\}\,. \tag{6.9.3}$$

Now this formulation has a very simple interpretation. It exhibits A and B as having no dependence on one another, nor on the settings of the remote polarizers (b and a respectively), but only on the local polarizers (a and b respectively) and on the past causes, c and λ. We can clearly refer to correlations which permit such factorization as "locally explicable". Very often such factorizability is taken as the starting point of the analysis. Here we have preferred to see it not as the *formulation* of "local causality", but as a consequence thereof.

6.10 Quantum mechanics can not be embedded in a locally causal theory

Let us define a correlation function $E(a, b, c)$ as the expectation value of the product of A and B:

$$E = \sum_{\lambda} \sum_{A,B} AB\ \{A, B|a, b, c, \lambda\}\ \{\lambda|a, b, c\}\,. \tag{6.10.1}$$

Here we have introduced a probability distribution $\{\lambda|a, b, c\}$ over the hypothetical complementary beables λ, for given values of the variables (a, b, c) which describe the experimental setup in the usual way. Now we make an important hypothesis:

> The variables a and b can be considered to be *free*, or *random*.

In the application to the Einstein–Podolsky–Rosen–Bohm two-photon experiment, a and b are the polarizer settings. Then we may imagine the experiment done on such a scale, with the two sides of the experiment separated by a distance of order light minutes, that we can imagine these settings being freely chosen at the last second by two different experimental physicists, or some other random devices. If these last second choices are truly free or random, they are not influenced by the variables λ. Then the resultant values for a and b do not give any information about λ. So the probability distribution over λ does not depend on a or b:

$$\{\lambda|a,b,c\} = \{\lambda|c\} . \tag{6.10.2}$$

We will come back to this. Then, using also the factorizability consequent on local causality,

$$E(a,b,c) = \sum_{\lambda}\sum_{A,B} AB\ \{A|a,c,\lambda\}\ \{B|b,c,\lambda\}\ \{\lambda \mid c\} . \tag{6.10.3}$$

From this it is a matter of simple manipulation to derive the

Clauser-Holt-Horne-Shimony-Inequality:

$$|E(a,b,c) - E(a,b',c)| + |E(a',b,c) + E(a',b',c)| < 2 . \tag{6.10.4}$$

But according to quantum mechanics, this expression can approach $2\sqrt{2}$. So quantum mechanics can *not* be embedded in a locally causal theory.

An essential element in the reasoning here is that a and b are free variables. One can envisage then theories in which there just *are* no free variables for the polarizer angles to be coupled to. In such "superdeterministic" theories the apparent free will of experimenters, and any other apparent randomness, would be illusory. Perhaps such a theory could be both locally causal and in agreement with quantum mechanical predictions. However I do not expect to see a serious theory of this kind. I would expect a serious theory to permit "deterministic chaos", or "pseudorandomness", for complicated subsystems (e.g. computers) which would provide variables sufficiently free for the purpose at hand. But I do not have a theorem about that [10].

6.11 But still, we can not signal faster than light

According to the above reasoning, the nonlocality of quantum mechanics can not be attributed to incompleteness, but is somehow irreducible. It remains however that *we* are very much bound by time and space, and in particular that we can not signal faster than light. Suppose that the two experimenters of the above were to try to communicate with one another by means of the apparatus in place. What could they do? We have supposed that one of them can freely manipulate the variable a, and the other the variable b. But each has to accept A or

B as it appears from his equipment, and neither knows the hidden variables λ. Now it is readily verified from the explicit quantum mechanical predictions for the EPRB gedankenexperiment that

$$\{A|a,b,c\} = \{A|a,c\}\,, \qquad \{B|a,b,c\} = \{B|b,c\}\,. \tag{6.11.1}$$

That is to say that, when averaged over the unknown λ, manipulation of b has no effect on the statistics of A, and manipulation of a has no effect on the statistics of B. And this is quite generally a consequence of "local commutativity" in so far as the variables a and b represent choices of "measurements", or "preparations", or "external fields".

6.12 Conclusion

The obvious definition of "local causality" does not work in quantum mechanics, and this cannot be attributed to the "incompleteness" of that theory [11].

Experimenters have looked to see if the relevant predictions of quantum mechanics are in fact true [1,2,9,12]. The consensus is that quantum mechanics works excellently, with no sign of an error of $\sqrt{2}$. It is often said then that experiment has decided against the locality inequality. Strictly speaking that is not so. The actual experiments depart too far from the ideal [13], and only after the various deficiencies are "corrected" by theoretical extrapolation do the actual experiments become critical. There is a school of thought [14] which stresses this fact, and advocates the idea that better experiments may contradict quantum mechanics and vindicate locality. I do not myself entertain that hope. I am too impressed by the quantitative success of quantum mechanics, for the experiments already done, to hope that it will fail for more nearly ideal ones.

Do we then have to fall back on "no signalling faster than light" as the expression of the fundamental causal structure of contemporary theoretical physics? That is hard for me to accept. For one thing we have lost the idea that correlations can be explained, or at least this idea awaits reformulation. More importantly, the "no signalling ..." notion rests on concepts which are desperately vague, or vaguely applicable. The assertion that "we cannot signal faster than light" immediately provokes the question:

Who do we think *we* are?

We who can make "measurements", *we* who can manipulate "external fields", *we* who can "signal" at all, even if not faster than light? Do *we* include chemists, or only physicists, plants, or only animals, pocket calculators, or only mainframe computers?

The unlikelihood of finding a sharp answer to this question reminds me of the relation of thermodynamics to fundamental theory. The more closely one looks at the fundamental laws of physics the less one sees of the laws of thermodynamics. The increase of entropy emerges only for large complicated systems, in an approximation depending on "largeness" and "complexity". Could it be that causal structure emerges only in something like a "thermodynamic" approximation, where the notions "measurement" and "external field" become legitimate approximations? Maybe that is part of the story, but I do not think it can be all. Local commutativity does not for me have a thermodynamic air about it. It is a challenge now to couple it with sharp internal concepts, rather than vague external ones. Perhaps there is already a hint of this in "quantum mechanics with spontaneous wavefunction collapse" [15,16]. But that is another story. As regards the present situation, I end here with Einstein's judgement, as translated by Casimir [23], on the new cookery of quantum mechanics:

> "...in my opinion it contains all the same a certain unpalatability."

Appendix: History

It would be interesting to know when and how the idea of the velocity of light as the limit developed. The earliest reference that I know is to a remark of G.F. FitzGerald, in a letter [17] of Feb. 4, 1889, to O. Heaviside. Heaviside had calculated the electromagnetic field of a uniformly moving rigid sphere. He did this at first for velocity less than that of light. Writing to FitzGerald he said that he did not yet know what happened for motion faster than light. FitzGerald remarked "... I wonder if it is possible ... ". Heaviside went on to solve the problem with velocity greater than c, and found that the solution is indeed rather different in character from that in the subluminal case. But he, at least at that time, saw no reason for not considering superluminal motion.

The idea of the velocity of light as the limit was one of the themes of Poincaré's famous address to the 1904 International Congress of Arts and Science at St.Louis [18]. After reviewing the experiments and ideas that we now see as leading up to special relativity theory, he said [19]:

> "...from all these results, if they were confirmed, would emerge an entirely new mechanics, which would be characterized by this fact that no velocity could exceed that of light any more than any temperature can fall below the absolute zero ... ".

One of the reasons that he gave for this was the increase of inertia with velocity [20]:

> "... perhaps we will have to construct a new mechanics, that we can only glimpse, where, inertia increasing with velocity, the velocity of light would become an uncrossable limit ...".

The advocates of "tachyons" have since pointed out that one can imagine particles which are *created* moving faster than light, without having to be accelerated up from a subluminal velocity. Poincaré also had another argument, concerning signalling and the regulation of clocks [21]:

> "... what would happen if one could communicate by signals whose velocity of propagation differed from that of light? If, after having synchronised clocks optically, one wished to verify the adjustment with the help of these new signals, one would find discrepancies which would show up the common motion of the two stations ...".

But in Switzerland you can set your watch by observing the trains go through the stations and looking up the timetable. Your watch is then synchronized with all the station clocks in Switzerland, and with the Federal Clock at Neuchatel. Although the trains do not go with the velocity of light, no discrepancies have ever been observed, and certainly none that would allow the detection of the motion of the stations, with the rest of Switzerland, through the aether[1]. The timetables allow for the finite propagation time of trains, but of course such allowance is necessary even with light. And clearly the same result will be obtained with any other method when proper allowance is made for the relevant laws of propagation, subluminal or superluminal, provided those laws are as regular as those of Swiss trains. I think that Poincaré nodded here. However he was not himself very convinced by his reasoning. Immediately after the last passage quoted he raises the possibility that gravitation goes faster than light. But a few pages later he is firmly maintaining that the motion of the stations will not be detected [22]:

> "... Michelson has shown us, as I have said, that the procedures of physics are powerless to show up absolute motion; I am convinced that it will be the same for astronomical procedures however far the precision is pushed."

[1]At still higher accuracy, even with light-signal synchronization, small cumulative discrepancies should appear. They would show up, not the mere motion of Switzerland through the aether, but that that motion is not just one of uniform translation, and that gravitation is at work, and that these affect even *Swiss* clocks.

References

[1] *International Symposium on the Foundations of Quantum Mechanics in the Light of New Technology 1, Tokyo 1983*, Physical Society of Japan, 1984.

[2] *International Symposium on the Foundations of Quantum Mechanics in the Light of New Technology 2, Tokyo 1986*, Physical Society of Japan, 1987.

[3] H.B.G. Casimir, *Haphazard reality*, Harper and Row, New York, 1983.

[4] H.B.G. Casimir, *Koninklijke Nederlandse Acadamie van Wetenschappen 1808-1958*, Noord-Hollandsche Uitgeversmij., A'dam, 1958, pp. 243–251.

[5] A. Einstein, Ann. Phys. **23** (1907) 371–384.

[6] A. Miller, *Albert Einstein's special theory of relativity*, Addison-Wesley, Reading, Mass., 1981, p. 238.

[7] M. Gell-Mann, M. Goldberger and W. Thirring, Physical Review **95** (1954) 1612.

[8] I think this was taken for granted by the early writers. It was spelled out by P. Eberhard, Nuovo Cimento **B46** (1978) 416–417.

[9] Many of the early papers are collected together by J.A. Wheeler and W.H. Zurek (editors), *Quantum theory and measurement*, Princeton University Press, Princeton N.J., 1983.

[10] This issue was raised briefly in a discussion among Bell, Clauser, Horne and Shimony, in 1976 in *Epistemological Letters*, reproduced in: Dialectica **39** (1985) 85-110.

[11] For a spectrum of recent views see J.T. Cushing and E. McMullin (editors), *Philosophical consequences of quantum theory*, Notre Dame, Indiana, 1989.

[12] *Quantum mechanics versus local realism*, ed. F. Selleri, Plenum Publishing Corporation, 1988. A.J. Duncan and H. Kleinpoppen review the experiments.

[13] J.S. Bell, *Speakable and unspeakable in quantum mechanics*, Cambridge University Press, 1987. An "ideal" experiment is sketched in paper 13 (Comments on Atomic and Molecular Physics **9** (1980) 121) of this collection.

[14] See for example the contributions of Ferrero, Marshall, Pascazio, Santos and Selleri, to Selleri [12].

[15] G.C. Ghirardi, A. Rimini and T. Weber, Physical Review **D34** (1986) 470.

[16] J.S. Bell, *Schrödinger, Centenary of a polymath*, ed. C. Kilmister, Cambridge University Press, 1986, reproduced as paper 22 in Bell [13].

[17] A.M. Bork on G.F. FitzGerald, *Dictionary of scientific biography*, Scribner, New York, 1981.

[18] *Physics for a new century*, ed. K.R. Sopka, Tomash Publishers, American Institute of Physics, 1986, p. 289.

[19] H. Poincaré, *La Valeur de la Science*, Flammarion, 1970, p. 138.

[20] H. Poincaré, *La Valeur de la Science*, Flammarion, 1970, p. 147.

[21] H. Poincaré, *La Valeur de la Science*, Flammarion, 1970, p. 134.

[22] H. Poincaré, *La Valeur de la Science*, Flammarion, 1970, p. 144.

[23] H.B.G. Casimir, in: *The lesson of quantum theory*, eds. J. de Boer, E. Dal and O. Ulfbeck, Elsevier, 1986, p. 19.

Part IV: New Pathways in Physics

Between Science and Technology
A. Sarlemijn and P. Kroes (Editors)
Elsevier Science Publishers B.V. (North-Holland), 1990

Chapter 7

Phase jumps – The physical basis of life

M. Eigen

Max-Planck Institute for biophysical chemistry

Göttingen

7.1 Introduction[1]

The bonding force between the natural world and the cultural world is man. Is it possible to extend the physical concept of phase changes from organised matter via life itself, to consciousness and into the domain of society and culture? Is there such a thing as an internally logical, or maybe even a "natural" preprogrammed consistency, which is more than an obvious analogy? In physics the term "phase change" is clearly defined and it is also applied to other temporal changes, usually changes which are continuous. Can the term be applied with the same precision to phenomena which transcend matter, phenomena which one could say are in principle immaterial in nature, although they assume the existence of matter?

The term "phase" (Greek: "phasis" usually translated as "appearance") is used in physics to define a certain state, limited not only in time but also in space. A simple example of

[1]This article is a translation of "Biologische Selbstorganisation; Eine Abfolge von Phasensprüngen", in: *Phasensprünge und Stetigkeit in der natürlichen und kulturellen Welt*, eds. K. Hierholzer and H.-G. Wittmann, Wissenschaftliche Verlagsgesellschaft mbH, Stuttgart, 1988, pp. 113–147. The editors thank the Wissenschaftliche Verlagsgesellschaft mbH for permission to publish this translation.

a continuous temporal change is harmonic oscillation. It can be represented as the regular circular motion of a point. The phase in this case is defined by the angle between zero and three hundred and sixty degrees according to the location of the point on the circumference of the circle. The term "phase" is also used in connection with complicated periodical motions. The phases of the moon are the result of the superposition of the orbit of the moon around the earth on that of the earth around the sun. The phases of biological cell division — pro-, meta-, ana- and telo-phases — demarcate time intervals within complex cycles of chemically controlled processes. Finally, the term is used for any, not necessarily periodic, temporal process. For example, one may speak of the creative phase of an artist.

On the other hand, the term "phase" also describes a spatially bound physical state. In this case, phase changes or reversals are discontinuities, jumps in the value of a parameter like internal energy or heat capacity. The three main material phases or states of matter — gas, liquid and solid — are well known. Physicists and chemists make much finer distinctions though: in the solid state they distinguish the crystal from the amorphous state, or the vitreous state, as an intermediate state between solid and liquid phases. The ferro- and the anti-ferromagnet represent typical material phases, ordered states of matter (fig. 7.1).

In as far as the physical state of a material system can be characterized by an order parameter, phase changes generally represent unsteady changes in order parameter, breaks in symmetry [2]. In this case it is not relevant whether the system is in equilibrium or far from equilibrium. A branch of physics, called synergetics by Hermann Haken [3], studies phenomena encompassing stable or metastable behaviour as well as instabilities, complex oscillations or so-called chaos [4].

Now let us pose the following question: is the physically well-defined concept of a phase change restricted to physical phenomena in the space-time-world, or can it be extended (strictly, and not as a mere analogy) to phenomena that transcend matter? Specifically, can we apply the concept of phase and its changes to ordered states of living matter, i.e., to ordered states that have a physical basis but where the spatial arrangement of the material substrate is not relevant any more? Can we apply this concept to a higher order of information, to a level manifesting self-consciousness, the essential hallmark of life, mind, society and culture? Structural complexity is the common characteristic of these states of order. Indeed, the degree of physical complexity that is manifested by these structures and modes of organization is so enormous that it seems a priori impossible to understand them physically. But we should realise that physics is not concerned with the complex structures and processes of

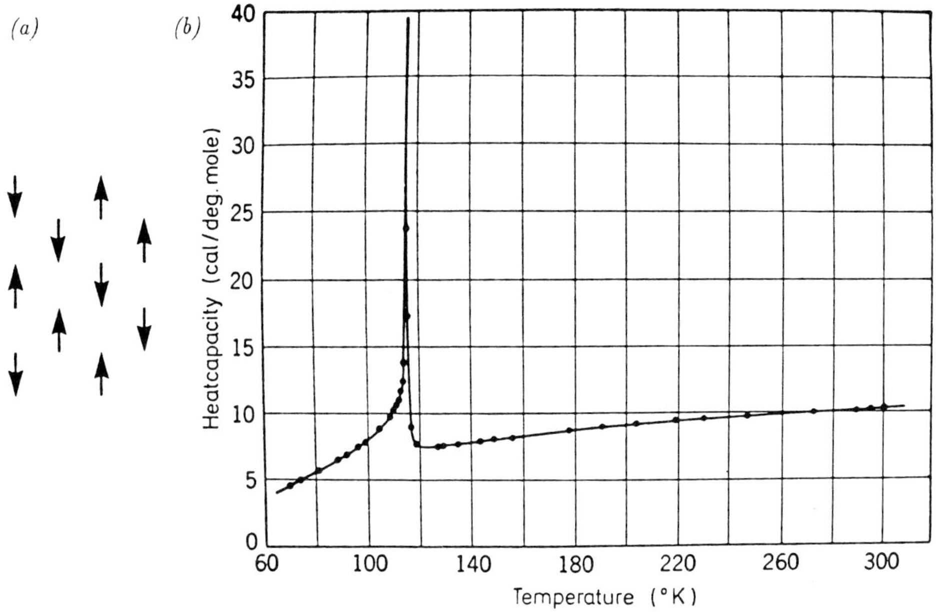

Figure 7.1: *(a) Spin direction of manganese ions in the antiferromagnetic crystal lattice of manganese oxide (MnO). (b) Heat capacity of manganese oxide (MnO) against temperature according to R.W. Millar [1].*

the real world but exclusively with the regularities in structures and the reproducible laws in processes. Although laws of formation, algorithms and forms of symmetries are often simple, the complex forms of appearance of the real world are usually not.

7.2 Simplex sigillum veri — Res complicatae sunt

Figures 7.2a to k show extremely complex — yet aesthetically very exciting — structures. What we see are computer-graphics that come from an experimental study performed by Heinz Otto Peitgen and Peter Richter and their colleagues in the Complex Dynamics research group at the University of Bremen [5].

The images have been "composed" with the help of a computer programmed with a very simple feedback algorithm. Feedback in this case means that a mathematical function is fed with its computed value as input over and over again. One starts with an initial value for x

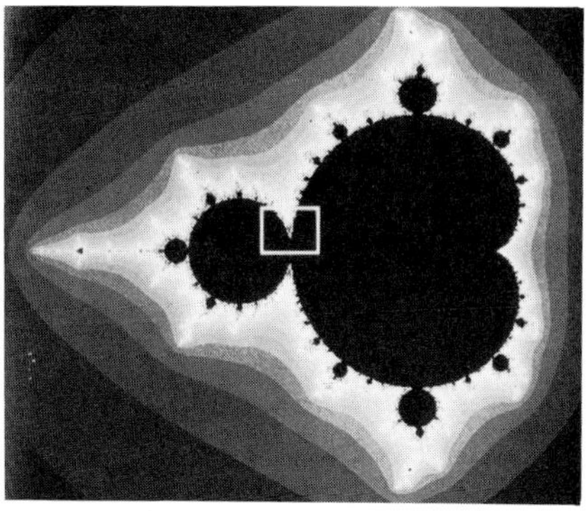

(a)

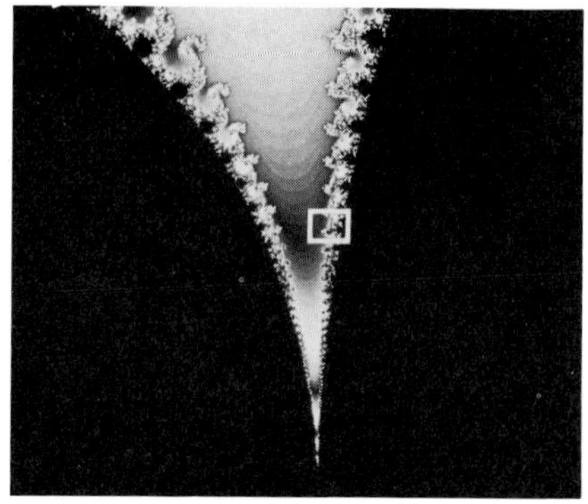

(b)

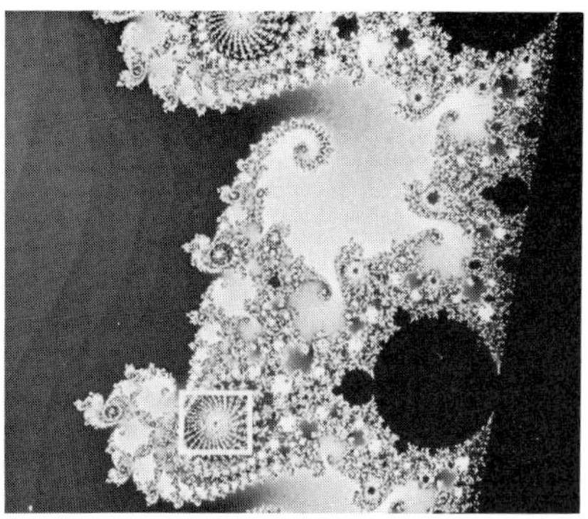

(c)

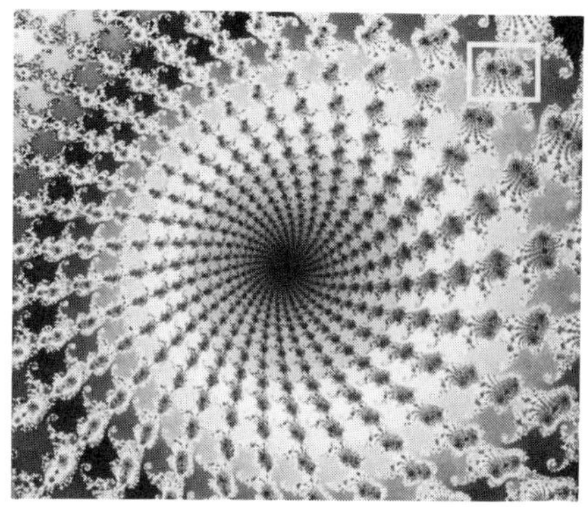

(d)

(e)

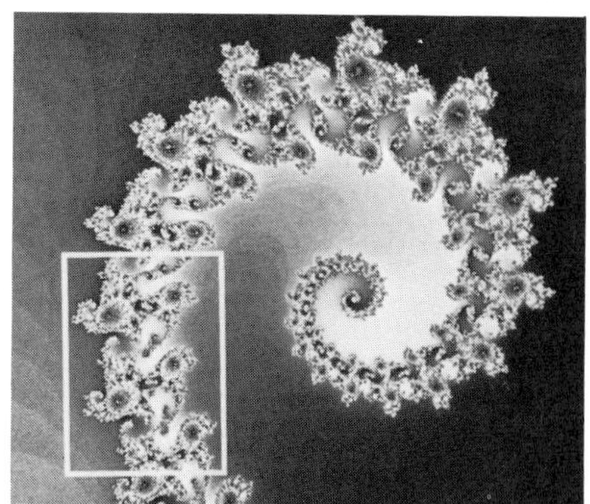

(f)

Figure 7.2: *The figures refer to the following range of the plane of complex numbers: on the real axis (left to right), values between -2.25 and +0.75; on the imaginary axis (bottom to top), values between -1.5 and +1.5. The figures (c) and (e) are identical; (d) and (f) show two corresponding enlarged sections.*

and feeds it into an equation — e.g. $x^2 + c$, where c is any given constant (which remains fixed during all steps of the computation). The result of the calculation is used as the new x-value and applied to the same equation. The images shown represent something like topographical charts in which the rapidity of the approach to infinity, as a result of reiterative calculation, is indicated through the use of different shades of grey for different x values. In this case the images were made by using the simple algorithm

$$x_0 \;\rightarrow\; x_0^2 + c = x_1 \;\rightarrow\; x_1^2 + c = x_2 \;\rightarrow\; \ldots\,, \tag{7.2.1}$$

however x and c were complex numbers. Each point on the topographical map corresponds to a complex number, the horizontal axis representing the value of the real part and the vertical axis the imaginary part. The origin of the coordinate system in figure 7.2a can be found within the pattern drawn in black; figures 7.2b to 7.2k show successively enlarged sections. The complexity of these patterns becomes clear: with each enlargement new details emerge, showing similar structures in which one can recognize all sorts of variations of the shape shown in figure 7.2a. Through scale enlargement one never reaches a limiting structure which is not continually changed by new detail.

It is simple to explain this behaviour with real numbers. The outcome depends decisively on the choice of the constant c. For $c = 0$, the behaviour is trivial. The solutions will go to infinity for any initial value $x > 1$ and will go to zero for any initial value $x < 1$. Therefore, zero and infinity are attractors, while the value 1 represents the borderline between the domains of different attractors. The results become interesting if one attributes the value -1 or -2 to c and starts the series with $x = 0.5$. In the first case the resulting values oscillate; in the second case, however, they seem to fluctuate irregularly as though they where produced by tossing dice. They are certainly not random; rather, they are deterministic solutions of the equation mentioned above. Despite this determinism, the starting value must be known very precisely to predict the answer after having gone through a long series of iterations. For the mathematician the starting value 0.5 is defined exactly. In reality, of course, it is virtually impossible to determine a numerical value, an experimentally measured value for example, with an arbitrary degree of exactness. Therefore, this peculiar and, in the long term, unpredictable behaviour is called deterministic chaos.

The above exercise may look futile; however, it has serious scientific use. The quadratic form (used in the equation) played an important role in understanding non-linear phenomena in physics. It returns in a different form in the dynamics of populations, in laser physics, in hydrodynamics (in describing turbulence) and in phenomena involved with molecular evolution as described in this article. Fundamental mathematical insights have been gained in the past years by analyzing the solutions, their domains of attraction, and borderlines between domains [5].

Especially striking is the beauty and aesthetic harmony of these pictures. They remind us of shapes that we may find in the living world, or of patterns that might spring from our imagination. Do natural phenomena arise from similar algorithms whose effects and manifestations are as unforeseeable as the chaotic solutions? Is there a connection between the two which goes deeper than just a formal analogy? We do not know! In the domain of life we find complex structures everywhere. However, complexity in the structures of reality does not necessarily mean complexity of the principles that shape reality. This is the lesson that we may draw from the figures shown above.

7.3 Algorithm of the origin of life

The complex appearances of living matter have also developed through a reiterative feedback principle that we call evolution. Evolution means: the better adapted form emerges from the more poorly adapted form. At the same time, it is generally true that a form distinguished by superior performance and adaptation is also represented by a more complex structure. Charles Darwin [6] formulated the algorithm of evolutionary adaptation. This natural algorithm is based on variation and selection. Ingo Rechenberg [7] used it for a new technical construction principle.

How did the transition "non living → living" come about? This is a difficult question, especially since it is difficult to define the "living" state. Let us consider a virus. Chemically speaking we are looking at a complex of molecules whose detailed composition is, in many cases, known exactly. Basically one could synthesize such a chemical substance in the laboratory. Like other homogenous chemical compounds, this substance can be crystallized and, in this form, it can hardly be distinguished from a mineral. However, when the substance is placed in a host cell, it then displays typical signs of a living organism: the virus reproduces itself; through mutations, it can adapt itself to the changing environment (and thus, for example, evade the immune response of the infected organism); and it undergoes a metabolism whose energy requirements are met by the host cell. In short: the virus avoids decline to thermodynamical equilibrium and keeps itself "alive"; it multiplies and adapts itself optimally to its environment; consequently, it evolves on the basis of natural selection. What is it then that distinguishes the living system from the non-living? Which property determines the difference between a chemical and a biological system? To answer this question, we must concentrate our interests on the chemical substance that is responsible for the principle of operation of the virus, namely the nucleic acids. The historical transition non-living → living must also have taken place with the appearance of these chemical substances.

The logic of life has its origin in physics and chemistry. Yet this logic took on an independent form at a very early stage.

The nucleic acids are situated at the border between chemistry and biology. Their specific chemical properties constitute the conditions for the transition from non-living matter to living matter.

The monomeric subunit of the nucleic acids, the nucleotide, is a defined chemical compound. It is made up of three sorts of compounds: a phosphate, a sugar (ribose or deoxyribose) and a heterocyclic base (purine or pyrimidine)(fig. 7.3).

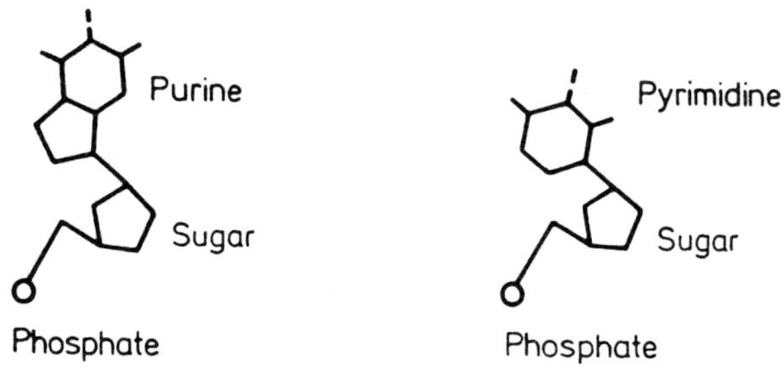

Figure 7.3.

If the monomeric components are in a high energy state (such as triphosphate for example) then they will combine spontaneously, without the help of enzymatic catalysts, to form macromolecular chains (fig. 7.4).

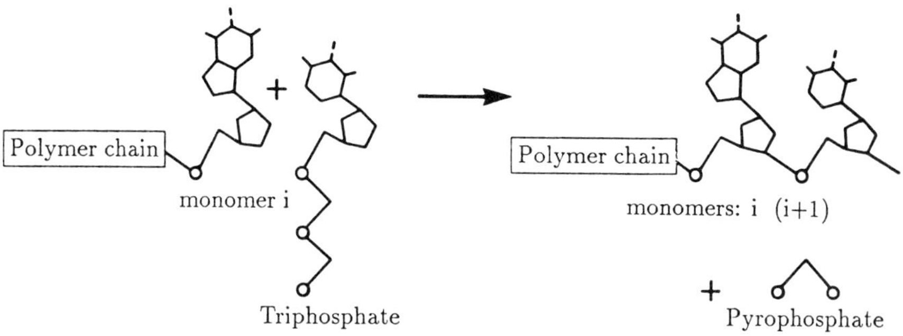

Figure 7.4.

Although we are still dealing with a typical chemical process, an essential requirement for selection and evolution has already been fulfilled. The nucleic acid chains so formed are metastable. They are spontaneously formed out of the high energy monomers and in an aqueous environment, they finally fall apart into low energy fragments. Formation and decay are irreversible processes. In the presence of water, the nucleic acids are not in chemical equilibrium. Selection and evolution require such a state, far from equilibrium, otherwise the system would not be able to unfold its characteristic behaviour which leads to optimal adaptation of all functions of life. In a living organism, metabolism has to produce a continuous supply of high energy components. Indeed, in evolution the nucleic acid building blocks (in the energy-rich triphosphate form, especially adenosine triphosphate (ATP)), have generally taken over the task of being the energy source of the cell. With the help of these energy rich components, every reaction is kept going; muscle activity in higher organisms is based on this principle too.

We are now at the crucial point that leads us from chemistry to biology: the four nucleotides bound in the macromolecule take on the role of symbols of a language. The sequence of these symbols can code a message. Thereby a completely new quality arises that does not occur in the physical-chemical conceptual framework, in which the notions of material interactions, of atoms, molecules or crystals, of energy forms and their transformations appear. This new quality is called information.

The definition of information demands: first, a limited number of symbols; second, the concatenation of these symbols to form chains or sentences whose structure can be laid down by a grammar and semantic conventions; and third, implicitly, a system to read (and if necessary to translate) the message contained in the sequence of symbols.

All the requirements for the use of nucleic acids as an information storage device are met by chemistry:

- The nucleic acids make use of four chemically classified components, two purines (A and G, short for adenine and guanine) and two pyrimidines (T and C, short for thymine and cytosine). The nucleotide T is only used in deoxyribonucleic acid (DNA). In ribonucleic acid (RNA), the "sister molecule" of DNA, which only differs from DNA with regard to a sugar (ribose instead of deoxyribose), the simpler pyrimidine uracil (U) is used instead of thymine.

- The concatenation of the nucleotides to macromolecules guarantees a metastable and, consequently, only a temporary existence of the message.

- There exists a specific interaction, mediated through hydrogen bonds, between each purine and one of the pyrimidines, such that A favours a place near T (or U) and G favours a place near C. This is called complementary base pairing. This complementary base pairing alone makes it possible to read the message at all and to save the temporarily stable information indefinitely by reproduction. Furthermore, this complementarity con-

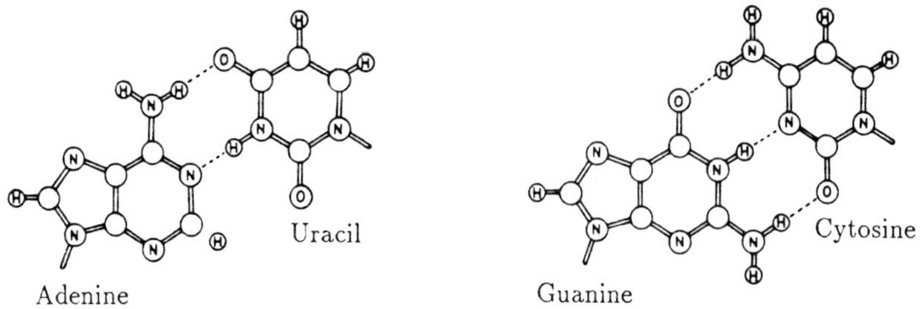

Figure 7.5.

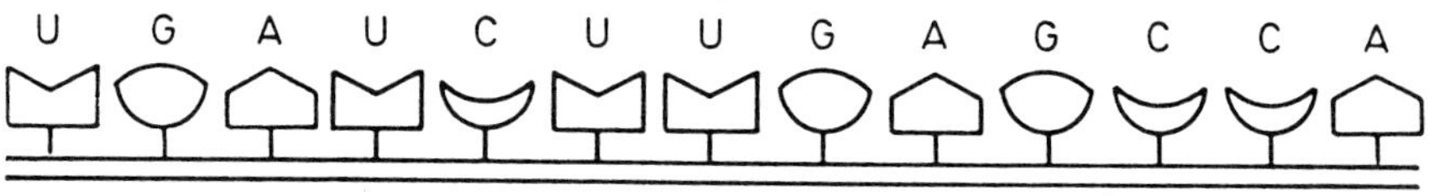

Figure 7.6.

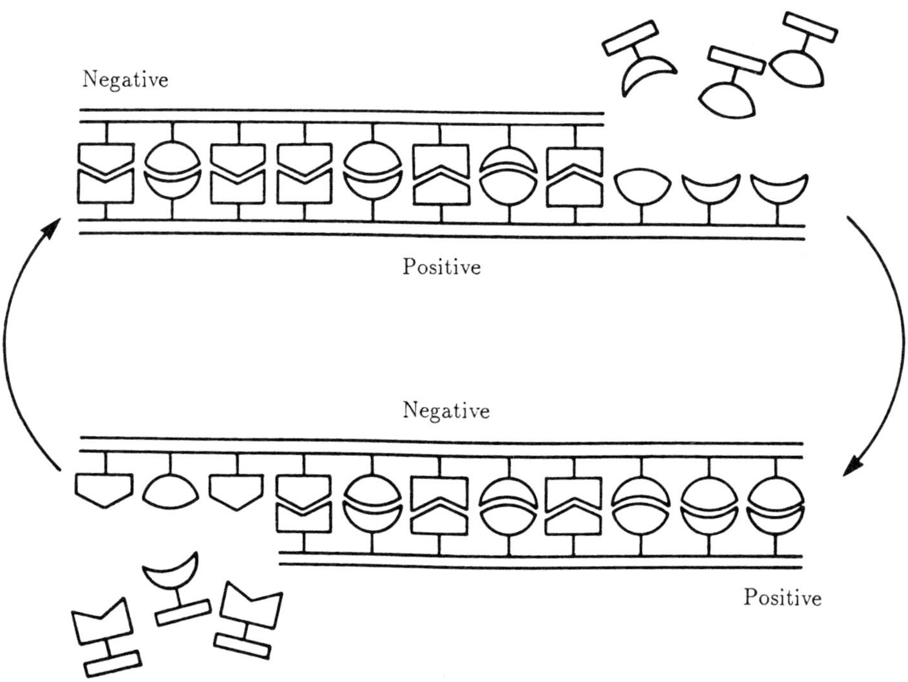

Figure 7.7.

veys recognition functions which are used in the translation of the genetic message as well as in the regulation and control of information processing (fig. 7.5).

Both base pairs have similar geometries but different bond strengths; the AU-pair, connected by two hydrogen bonds, is less stable than the GC-pair which is connected by three hydrogen bonds. If we represent the symbols in a simplified way (fig 7.6): we can (as shown in figure 7.7) characterize the process of replication; replication proceeds — in a fashion similar to photography — by means of a negative as the intermediate stage.

One must bear in mind that replication does not take place without errors because the complementary interaction is of finite strength and can be disturbed by the thermal motion of the molecules and also by chemical fluctuations. Thus, through the controlled operation of replication enzymes, error-copies or mutants are created which form the reservoir for the changes which the process of evolution requires. Self-reproduction, mutageneity, and metabolism, all characteristics of life grounded in the metastability of the chain of molecules, appear together on the evolutionary scene with the advent of nucleic acids; indeed, they are closely bound to this class of molecules.

Information is a property that transcends chemistry, a property that is typical of biology. The physical carrier of the information, the nucleic acid molecule, is unstable like any other chemical substance; for example, in an aqueous medium, it disintegrates through hydrolysis. In the living cell, this disintegration is regulated by enzymes. But the genetic message contained in the nucleic acid molecule is stable and, through self-reproduction, has become immortal. The information stored in our genes was produced, in principle, about three and a half to four billion years ago, namely at the time when nucleic acids began storing and passing on information, continually changing it, and transferring it in a modified form. During these processes, the material carrier had only a limited life span, only the immaterial quality information survived. This development can nowadays be traced back to the time when the genetic code emerged; the code can be dated quantitatively by comparing the information laid down in the different sequences of nucleic acids [8].

The chemical complementarity of nucleic acid building blocks constitutes the foundation for the new property information. It explains how information can be saved longer than the material lifespan and how communication is possible. On the other hand, the chemical property of complementarity does not say anything about the immaterial property information, just as the notes of a Mozart symphony captured in printer's ink have nothing to do with the quality of the music.

How is information created? We will see that once again it is possible to make a causal link between self-replication based on the complementary base pairing and the emergence of genetic information. It will become clear that the evolution of this immaterial property that we call information takes place in phase changes, and that we may describe this process, like other physical processes, in quantitative relations. At the same time, we will see that we can derive some sort of an understanding of the concept of information only through these relations, in the same way that we came to an understanding of the physical qualities matter, fields, and energy by studying their relationships.

To begin with, we must introduce a new concept of space borrowed from information theory. "Classical" information theory is concerned with the processing and communication of information, but not with the question of how information is created. Indeed, our deliberations will be based on information theory but will eventually go beyond it.

7.4 The concept of sequence space

What we are looking for is a suitable space into which we can map the nucleic acid sequences and describe their kinship relations correctly. Neighbouring sequences should be related to neighbouring coordinates, and this not only with regard to a given sequence such as the wildtype, but the mapping should also be valid for the mutual distances of all mutants. This means that one-, two- or three-dimensional mappings are excluded, for instance those in which the wildtype corresponds to the origin of the coordination system and the k-error- mutants to circles with radius k around the centre. The concept of sequence space can only be realized with the help of a multi-dimensional space. The scheme illustrated in figures 7.8 and 7.9 shows this. In this scheme, the sequence space for binary symbol series is constructed in an iterative manner.

To begin with, we consider just one position of the sequence; it can be occupied in two alternative ways. The sequence space consists of just two points which correspond to the two alternative possibilities: R (purine) or Y (pyrimidine). We connect neighbouring points (points that can be arrived at by a single mutation) by a straight line. In the case of two positions, the straight lines must be doubled because the added position itself has two ways in which it can be occupied. We then connect the corresponding end points of the two lines. Thus we obtain a quadratic surface. With three positions, this surface is doubled and extended to a cube. Clearly, one recognizes the iterative principle by which the scheme is constructed:

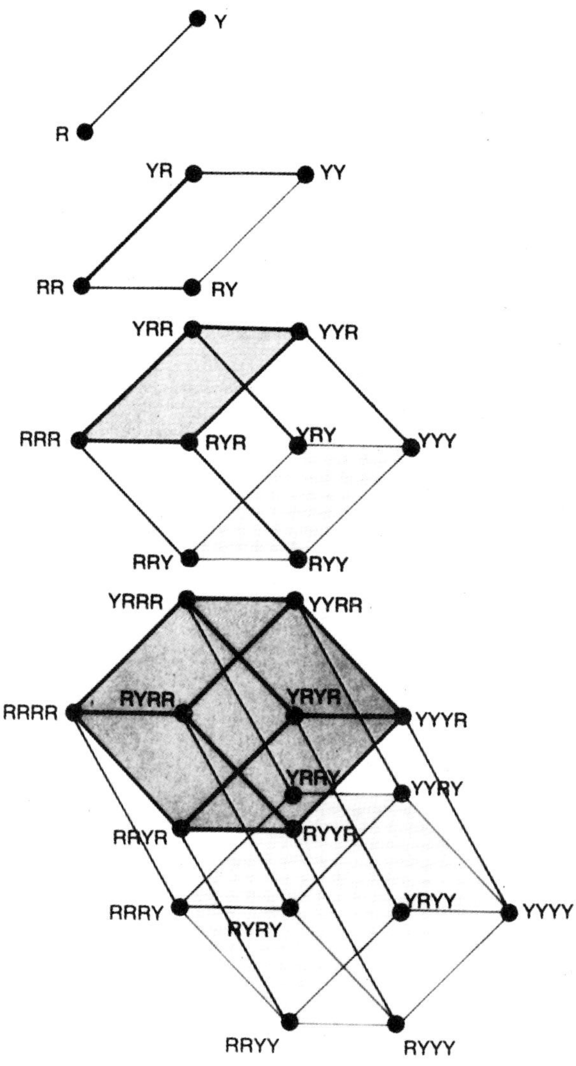

Figure 7.8: *Iterative construction of a binary sequence space:* $\nu = 1, 2, 3, 4$.

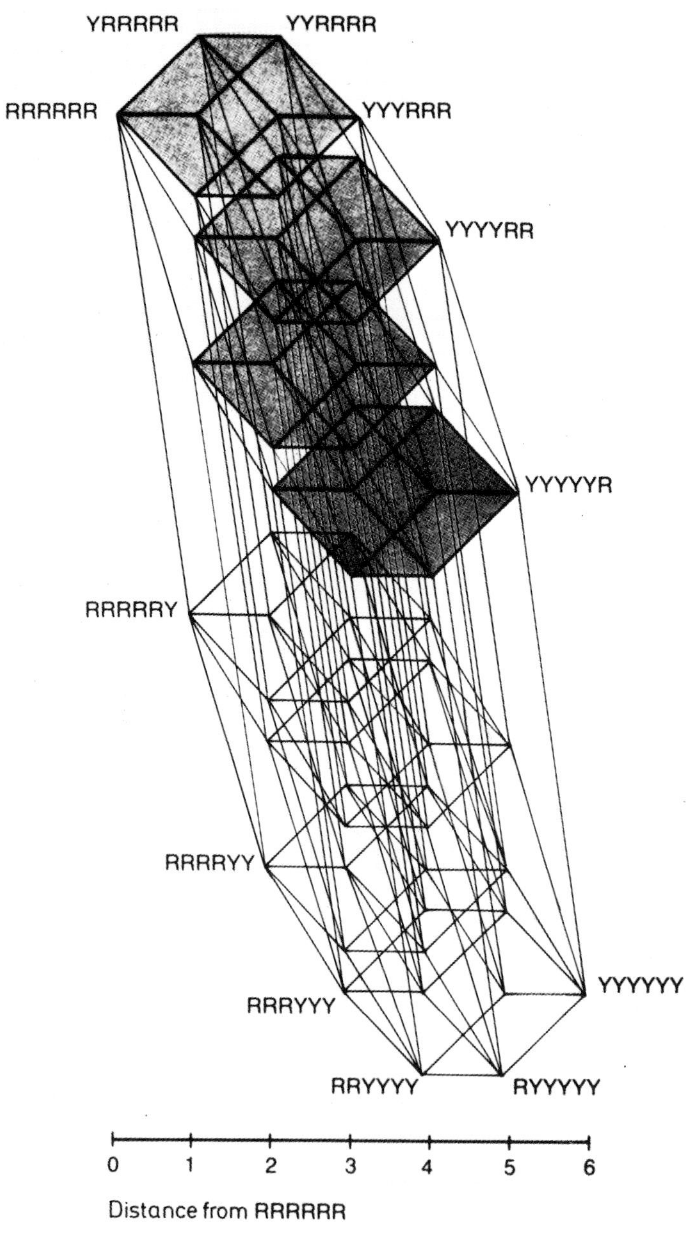

Figure 7.9: *Hypercube for* $\nu = 6$.

with each additional position the previous diagram is doubled and the corresponding positions of each diagram must be connected. In this way, we obtain a hypercube with ν dimensions, where ν is the number of positions in the sequence. The concept of sequence space was developed in information theory. It was introduced mainly by Richard W. Hamming [9]. Ingo Rechenberg [7] was the first to propose the application of this concept to nucleic acids. In a recently published monograph [17], the quantitative development of this concept, in relation to the evolution of nucleic acids [10–14] as well as to comparative sequence analysis [15,16], was described in detail.

Although we can hardly visualize a multidimensional structure, we can still deduce intuitively from the chosen representation of the six-dimensional hypercube (fig. 7.9) some characteristic properties.

To begin with, one notices that the hypercube has an enormous capacity which increases exponentially with the number of positions. Every additional position doubles the volume. Our universe, or better still, a sphere with a radius of ten billion light-years, contains the volume of approximately 10^{108} hydrogen atoms. (For simplicity's sake, we have taken the volume of a hydrogen atom to be one cubic Ångstrom.) If we assign one hydrogen atom to each point in our sequence space, then the hypercube of a sequence containing 360 symbols could harbour all the 10^{108} hydrogen atoms ($2^{360} \approx 10^{108}$). In other words, the volume of a universe in terms of Ångstrom cubes corresponds to the volume of the hypercube of a binary sequence of 360 positions.

If we ask someone the way in New York, he or she usually expresses distances in terms of the number of "street blocks". Similarly, we define the mutation distance between sequences as a number, namely the number of positions which are occupied differently in the two sequences that are compared. The fact that distances in the sequence space are very small, has an important consequence. Every destination can be reached very quickly assuming that one can orient oneself. Without such an ability we would surely be helplessly lost, as in the unfathomable spaces of the universe.

The degree of inter-linking is immense, as one can see immediately in the diagram. Only directly neighbouring points have been connected by lines in this diagram. As soon as one includes mutation jumps, i.e. changes of more than one position within one round of replication, the number of states that can be reached from each point (i.e. the number of connecting lines that originate from each point) soon becomes greater than any realizable population number.

All the characteristics mentioned are present even more markedly in the case of a sequence

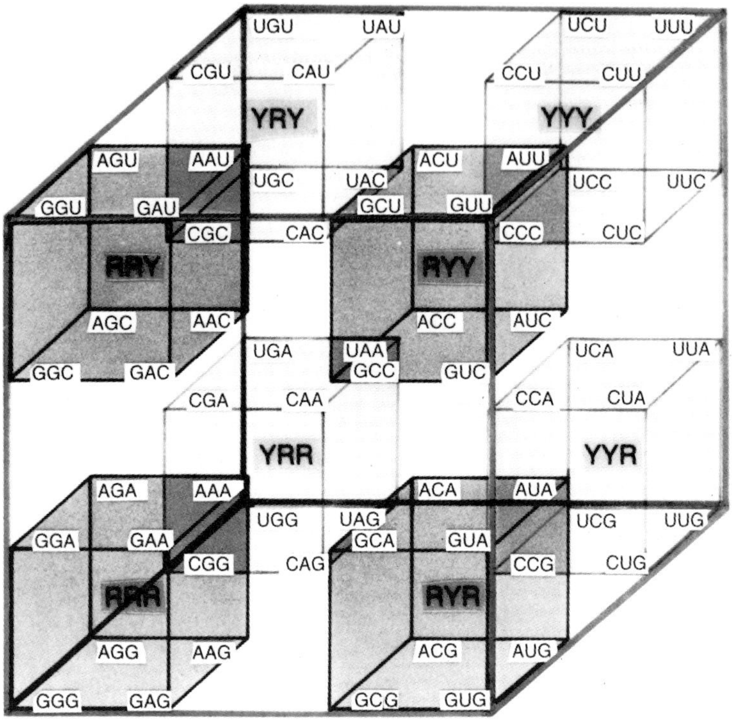

Figure 7.10: *Quaternary sequence space for $\nu = 3$.*

space which accounts for four kinds of symbols, as is materialized in nucleic acids. How can one visualize such a space? It is best to start with a binary sequence space and assign only the base classes R and Y to each position of the sequence. Each point in this space now corresponds to a certain sequence in RY-notation. One then allocates the precise base to each base class. That implies another binary decision for each position because there are two possibilities for both R (A and G) and Y (U and C). In other words: a ν-dimensional subspace is associated with each point in the ν-dimensional binary sequence space. At this stage the dimension is 2ν and we are now dealing with $2^{2\nu} = 4^{\nu}$ possible states. Figure 7.10 shows such a sequence space for $\nu = 3$. The scheme illustrates a perspective of the genetic code.

7.5 The value topography of the sequence space

The self-organization of a system of sequences can now be treated as a problem of occupation in sequence space. Assuming there is a tendency to favour one sequence, then all existing

Figure 7.11: *Fractal country (photographed by David Usher in Anga Borrego).*

sequences must be grouped at one single point in sequence space or in its immediate vicinity. This would correspond to the creation of information. On the other hand, if there were no tendency to prefer a sequence, then the existing sequences would be distributed over the sequence space, like molecules of an ideal gas in physical space, with the occupation probability of $4^{-\nu}$ (ν = sequence length) for each point. As an analogy of condensed information in sequence space, we might consider the distribution of water on the earth's surface. It gathers in the deepest places. If the earth were completely flat, then the water could not collect at any particular point on the surface; in this case the oceans would cover the earth evenly. In sequence space this would mean a very low probability for any point to be occupied.

The distribution of height on the surface of the earth corresponds to a distribution of selection values in the sequence space. Dynamic parameters are associated with each sequence, and these tell us how precisely and how fast the sequence in question reproduces itself and how long it lives. The selection value is a measure of the efficiency of multiplication. It indicates how many exact copies are made during the lifespan of the sequence. The complementarity of the chemical interaction between the bases of nucleic acids is fundamental to their ability to reproduce and determines the intrinsic autocatalytic nature of the synthetic processes which, in the absence of restrictions on growth, would lead to exponential multiplication. In terms of sequence space this means that the points with highest local selection value will be occupied most frequently. What does such a selection value topography in sequence space look like?

Let us once again use the comparison with the earth's surface. Benoit Mandelbrot [18] has shown that the structuring of a mountain can be represented by a fractal height distribution. Fractal structures are structures that look similar in all scales. See figure 7.2 for illustrations. Figure 7.11 shows us a typical height distribution of a mountain area (Anga Borrego).

Mandelbrot considers a process that one could regard as being the reverse of an evolutionary problem. He asked himself: What chance does a raindrop have to reach the ocean when it falls on an island with a height distribution as shown in figure 7.12? If the island, as in our projection, were actually one dimensional then all the troughs would have to fill up before the raindrop could flow, as over a terrace, into the ocean.

In a similar way, an upwardly directed evolution process would only occur in the value landscape if this process could smooth the "gendarmes". (A gendarme is a block-like elevation on a mountain ridge). However, the surface of the earth is not one-dimensional. The addition of a second dimension turns each trough into a real two-dimensional dip, whose edges have differing heights. The terrace figure shows us an arbitrary one-dimensional cross-section, so,

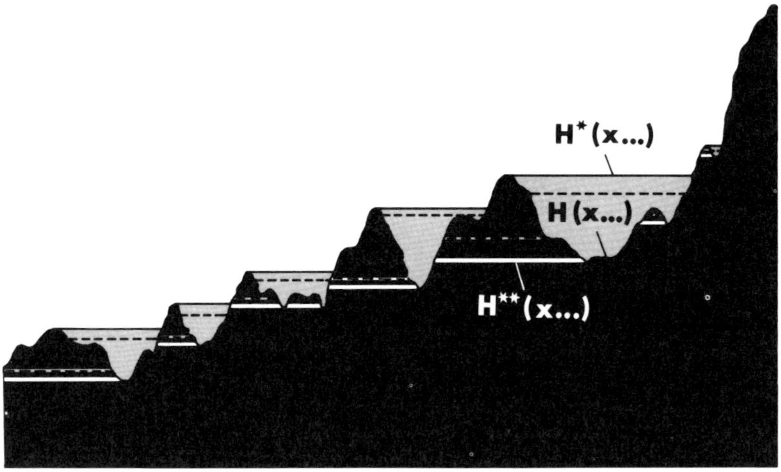

Figure 7.12: *One-dimensional height distribution.*

in this case, each maximum is merely a point on the edge of the trough, whose other points lie either higher or lower. Only in an extreme case does the one-dimensional cross-section go through the deepest point on the real edge of the trough, as shown in cross-section 3 of figure 7.13.

On the other hand, if there are lower points on the trough's edge — as generally is the case — then water would run off at those points. Consequently, one does not need to fill the trough to its highest point on the one dimensional cross-section because the water will drain off at a lower spot not covered by the one dimensional cross-section. Consequently, the principle implies that, with each additional dimension, the water level will drop systematically so that eventually the raindrop will always find its way to the ocean (as long as there are enough dimensions).

A reversal of the process leads to a similar situation: the evolutionary tour to the summit. Bear in mind that the sequence space is grained, that is, it represents a point-space. In this case, the height values change discretely and not continuously. Furthermore, mutation-jumps are possible, so that, from a given position, possibly 10^{10} or more neighbouring points may be reached by jumps. Thus the process of evolution in high dimensional sequence space can reach, with high probability, the summit of the value landscape through a quasi-continuous route.

While trying to visualize the multi dimensional value landscape, one must realize that one is never further away than ν steps from the highest summit. Because multiple connecting lines

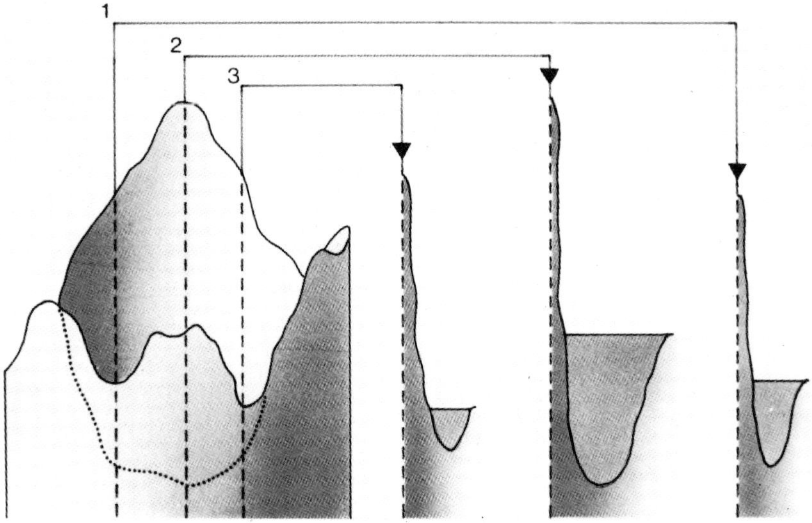

Figure 7.13: *Two-dimensional height distribution and three one-dimensional cross-sections.*

are possible — corresponding to the many possible mutation jumps — the probability that one will find the edge leading to the highest summit is much greater than it would be through a purely random walk.

7.6 Phase formation in sequence space — The concept of quasi-species

In the following, we will have a look at three concrete examples of value landscapes and we will calculate, using the methods of quantitative theory [14], what the corresponding occupation in the sequence space looks like.

Let us take a binary sequence with $\nu = 50$ positions and with a value distribution as shown in figure 7.14. The index i (= Hamming distance) indicates in how many positions the mutant differs from the wildtype. ($i = 50$ means that all 50 positions are occupied in a different way). The sequence with the index 0 (zero mutations) represents the wildtype, which is ten times "better" than all mutants (for example, because it reproduces itself 10 times faster than the mutants). All mutants ($i > 0$) have the same value of selection. Note that in total there are $2^\nu \approx 10^{15}$ different mutants, including the wildtype (0 errors), one antipode (50 errors) and among these also two sets of 50 different mutants which are 1-error or 49-error copies of the wildtype. The maximum number of possible error variants is found for $i = 25$. This set constitutes approximately 11 percent of the total number of mutants.

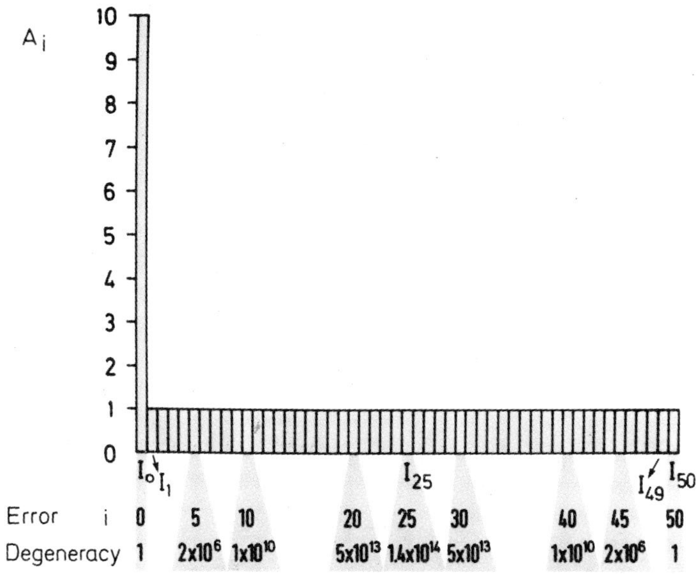

Figure 7.14: *Distribution of selection values A of i-error mutants for* $i = 0, 1, \ldots, 50$.

We now calculate the relative population number (x), summed over all individual i-error mutants, in relation to the error rate $1 - q$. x is the number of copies corresponding to the Hamming distance i, divided by the total number of copies. The probability of copying a symbol correctly is q. Consequently the error rate per symbol is $1 - q$. The calculations were done by Peter Schuster and Jörg Swetina [19]. The top diagram in figure 7.15 shows the relative portion of wildtype ($i = 0$) and the number of error-copies for $i = 1, 2, \ldots, 50$. In the bottom part, the relative population number of the wildtype x is shown logarithmically. One sees that only when the error rate is zero a clear "all or nothing" decision (survival of the fittest) is at hand. With increasing error rate, the population number of the wildtype quickly decreases to relatively low values such that soon there are in total more one-, two-, and three error-mutants, etc. than wildtype. Despite this, the wildtype is selected in a stable way. The mutants are distributed around the wildtype and its sequence is clearly defined as the consensus sequence of the distribution. Only when one reaches the error threshold (see figure 7.15 at $1 - q = 0.046$) does the information get lost. The logarithmic representation clearly demonstrates this effect. At the error threshold, the population number of the wildtype drops sharply from several percent to about 10^{-15}. This (i.e. 2^{-50}) is the value one would expect for a sequence length of fifty binary symbols, if there is no preference. Thus, the wildtype sequence is just one of the $2^{-50} \approx 10^{-15}$ possible variations. Then the one-error mutants, of which there are fifty different

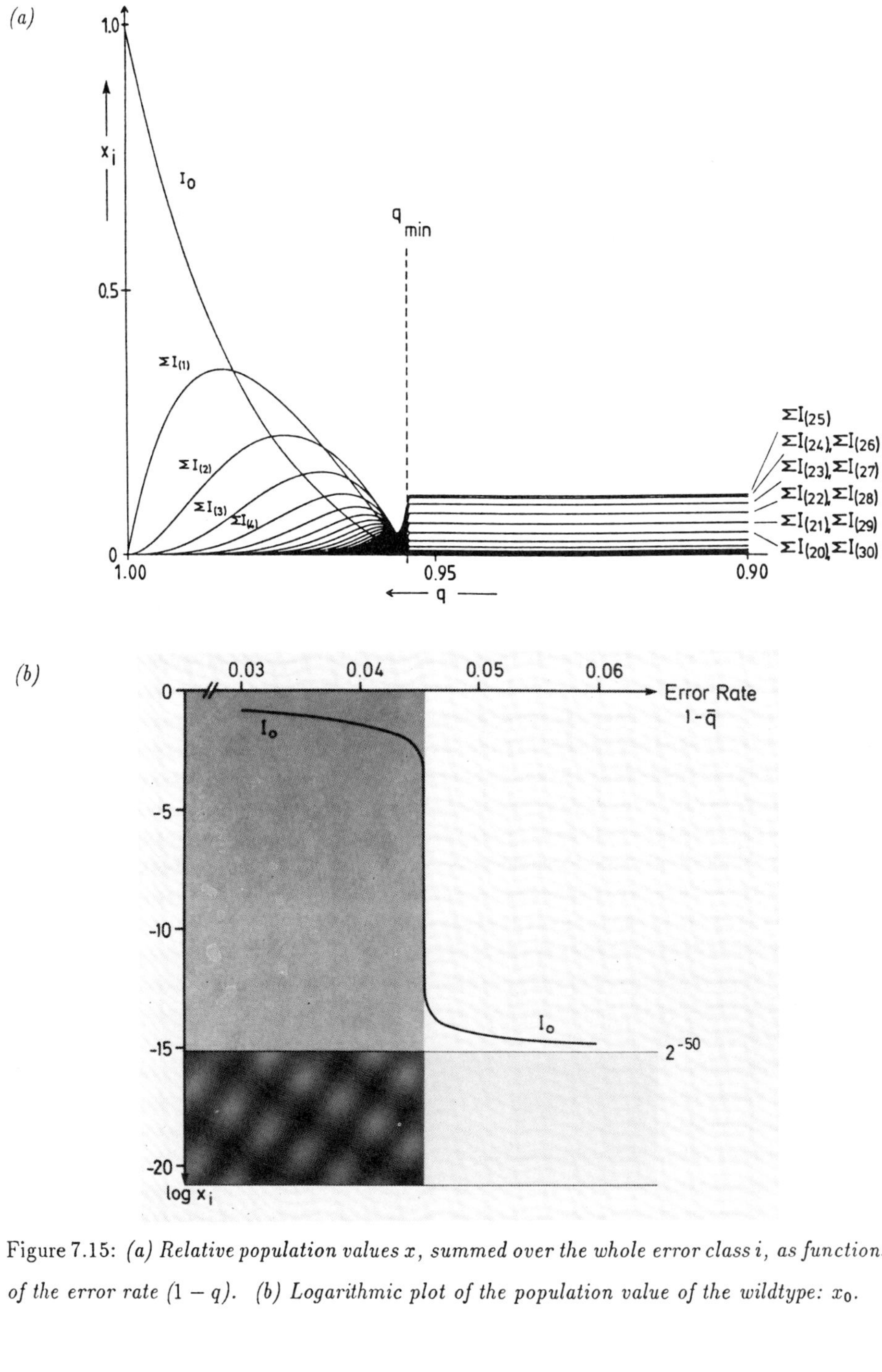

Figure 7.15: *(a) Relative population values x, summed over the whole error class i, as functions of the error rate $(1-q)$. (b) Logarithmic plot of the population value of the wildtype: x_0.*

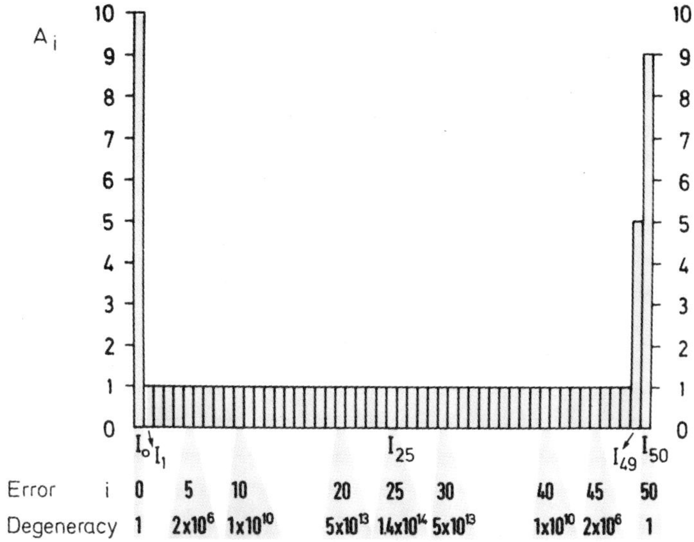

Figure 7.16: *Selection value distribution A of the i-error mutants for $i = 0, 1, \ldots, 50$.*

types, appear 50 times more frequently. Correspondingly all other mutants are populated according to their degeneracy, up to the 25-error mutants, which are most prevalent. Each individual mutant has the same expectation value (i.e. $2^{-50} \approx 10^{-15}$). This means that all information is irretrievably lost.

The process looks like a phase transition. The threshold value corresponds to a melting temperature. Beyond the critical threshold, the information "melts". It gets lost due to an accumulation of errors. However, the best conditions for the evolution are found close to the critical error threshold. In this region, the wildtype is still stable as long as no better adapted variant shows up in the spectrum of mutants. At the same time, the maximum number of possible mutants is generated, and amongst these mutants are some whose distance from the wildtype is relatively large. In a changing environment, such a system can adapt itself very quickly.

Experiments with viruses show that natural mutant ensembles actually operate directly at the error threshold. Systems with a correspondingly adapted error rate apparently have a clear advantage in the process of evolution. The threshold value of the error rate is inversely proportional to the quantity of information and sequence length respectively. The influenza virus, which has about 14.000 nucleotides, replicates with an error rate just smaller than 10^{-4}. When trying to determine the exact value of the threshold error rate, one must take into

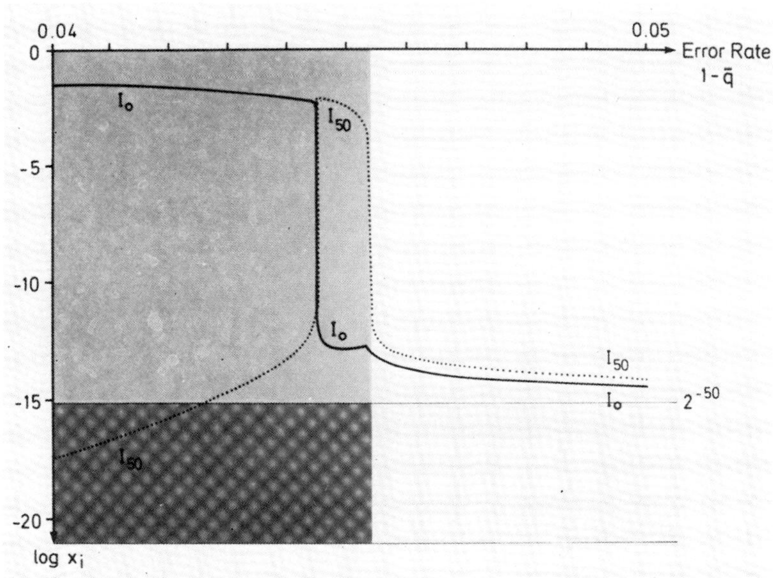

Figure 7.17: *Relative population numbers x, summed over the whole error class i, represented logarithmically as a function of the error rate $(1 - q)$.*

account how much better the wildtype is adapted than its mutants. Neutral mutants are co-selected.

For the evaluation associated with selection, the value distribution, i.e., the value topography, in the entire sequence space plays a role. This becomes clear if one considers the second example in figure 7.16 [14,17,19]. In contrast to the previous value distribution, the 50-error mutant, the antipode of the wildtype, now has a selection value which lies only ten percent below that of the wildtype. Besides that, a selection value which is only half that of the wildtype is accorded to the fifty 49-error mutants (the 1-error mutants of the antipode of the wildtype). As before, the other mutants get a selection value which is one tenth that of the wildtype.

The diagram in figure 7.17 shows us the corresponding population distribution logarithmically. For the error rate $(1 - q) = 0$, a clear selection in the Darwinian sense again arises: survival of the fittest. But..., at an error rate between four and five percent, a phase transition occurs. Here, the 50-error copy, which in comparison to the wildtype is somewhat less adapted, proves itself to be superior thanks to the better value topography in its neighbourhood. At the transition point, because of the value topography of their respective neighbourhoods, the

0-error and the 50-error copies are exact neutral mutants, while above the transition point the 50-error copy is superior. Finally, with the further increase of $(1 - q)$ an absolute error threshold is reached. Beyond this threshold value, the system no longer contains any original information.

The theory shows that the "almost-neutral" mutants in particular, i.e., those mutants whose degree of fitness is almost equal to that of the wildtype, have a special influence. The third example in figure 7.18 makes this clear. In the top part of the figure, two value landscapes are shown which extend — starting from the wildtype (left-hand border of figure) — to a Hamming distance $i = 12$. One landscape (white bars) corresponds to the example in figure 7.14. All mutants are (at least) ten times less adapted than the wildtype. On the other hand, the second value distribution (dark bars) represents a typical mountain landscape. With increasing Hamming-distance, the reproduction rate decreases first to half its value (relative to the wildtype) and then again approaches almost the value of the wildtype. (The simulated landscape characterizes very realistically the situation in natural virus populations). In the lower part of the figure, the corresponding stationary population numbers are represented, this time for individual mutants.

What is surprising about this simulation is not so much the qualitative but the quantitative outcome. The decrease of the individual population numbers by a factor 10^{-20} in the first case is in agreement with expectation based on the example of figure 7.14. A 12-error mutant, in this case, would have no chance of ever appearing in a laboratory experiment in which one usually deals with about 10^{-10} to 10^{-12} particles.

The behaviour of the system is completely different in the mountain landscape: the population number of the mutants decreases evermore slowly with increasing distance from the wildtype; indeed, it finally becomes constant and increases gradually again. The emergence of the 12-error mutant is predictable and inevitable. It would appear deterministically in every laboratory experiment.

The difference in expectation values of the population numbers for both distributions is amazing. This difference in this case turns out to be about eighteen orders of magnitude and is based on the amplifying effect of a multiplicative relation: the 12-error mutant arises from the 11-error mutant, this one from the 10-error mutant, etc. From this we see that the probability of a certain mutant appearing can be steered. The production of (relatively well adapted) mutants along an edge of the value landscape takes place at a much higher probability than in a flat value region. Because edges lead to mountain peaks, the process of evolution proceeds

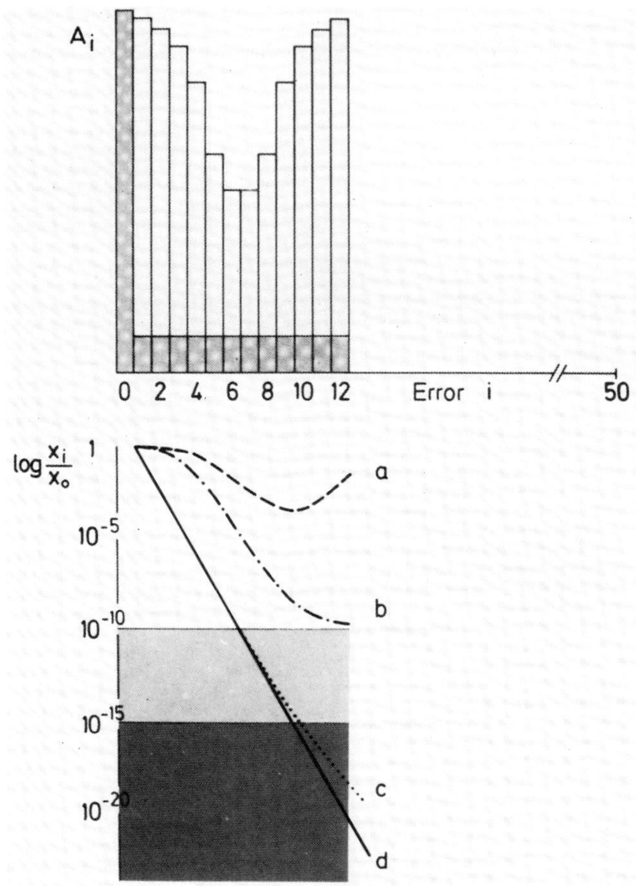

Figure 7.18: *(top) Two examples of selection value profiles (A) of a wildtype sequence* $(i = 0)$ *and its i-error mutants* $(i = 1$ *to* $12)$. *(bottom) Population numbers for individual mutants relative to the wildtype* (x_i/x_0) *represented logarithmically for both the selection profiles shown above (sequence length* $\nu = 50$, *error rate* $(1 - q = 1/\nu)$*). The curves a and b correspond to the mountain profile (light grey), curves c and d to the profile of the flat plane (dark grey). For curves a and c, we assume that all precursors of the 12-error mutant have the selection value profile shown; whereas in both curves b and d, only one precursor has the indicated selection value and all other values have been set to zero (singular mountain ridge).*

in a goal-directed fashion by virtue of a self-steering mechanism. Here one recognizes most clearly the difference between this process and the classical Darwinian concept of selection and evolution based purely on chance. For large numbers, the quantitative effect changes into a novel quality.

7.7 Darwin is dead — Long live Darwin!

The results of our deliberations so far need to be summarized in words and illustrated with experimental examples. Of course, one must keep in mind that the quantitative calculations are responsible for revealing the qualitative new view of the problem of evolution.

In this new view, evolution is still based on mutation and selection, chance and law. The phase jump in our knowledge is evident only at the level of detail. This becomes clear when we compare the new insights with the claims of the classical interpretation of Darwin's world view.

The target of selection, in the classical view, is the wildtype, which under normal conditions is identical to the type that has adapted itself most successfully. In case of incomplete adaptation, the evolutionary rate of progression must be dependent on the population size, simply because in a large population, favourable mutants appear statistically more frequently than in a smaller population. Furthermore, one assumes that each mutant owes its existence to an accidental event. The most favourable mutant is not produced on purpose and, consequently, can appear only very rarely.

Modern sequence analysis, determination of the sequence of the molecular components in a gene (or in its translation product, the protein), enables the testing of this conclusion. The result is, that in phylogeny the components of the genes are substituted at a constant rate in time which is, by and large, independent of the population size involved.

A certain uneasiness with the classical theory led some years ago to the development of the so called non-Darwinian neutral theory by Motoo Kimura [20]. This uneasiness was based on probability assessments which made it seem almost impossible that a structure as complex as the genome of a living organism could be built by trial and error sampling of all the possible mutations. Such a process would soon have to come to a halt at a hill of relatively low height on the value landscape. The neutral theory says that, besides the many favourable and the many more unfavourable mutants, there must also be many neutral (neither favourable nor unfavourable) mutants [21]. These would be able to grow, enabling a drift through the sequence space which would make it easier to find a favourable value-region. This process is independent of the population size. The frequency of generating a neutral mutant as such is, of course, proportional to the population size. On the other hand, the probability that this neutral mutant proliferates and displaces the previously established distribution — of equal value — is inversely proportional to the population size. The population size thus cancels out. The macroscopic emergence of the neutral mutant is, in the neutral theory, independent of the population size, and this interpretation agrees exceptionally well with experimental findings.

The theory of quasi-species emphasizes even more strongly the significance of neutral mutants and — in addition — of all those that are capable of reproduction. The influence of these mutants becomes stronger the less they differ in their selection value from the wildtype. It is the whole distribution of genetic sequences rather than an individual type which is subjected to an evaluation by selection. The distribution is ordered according to the mutation distances and the value topography in sequence space has a marked effect on the occupation numbers. Mathematically, the maximum eigenvalue of the value matrix to which all individual selection values contribute, rather than the selection value of an individual type, is decisive for the phenomenon of selection. Physically speaking, the phase behaviour described above is involved. The mutant colony is localized in the sequence space; the defined wildtype sequence which is available for analysis is a consensus sequence. It describes the centre of gravity of the distribution, the centre of localization of the population in the sequence space. Evolution is a series of phase transitions. In the process of evolution, the centre of gravity of the selected distributions within the sequence space shifts. This process follows a more or less coherent route.

One may object that all of this, and the neutral theory as well, is merely a description of detail, and besides that, a confirmation of the classical Darwinian view which offers no new fundamental insights. The situation can most easily be assessed with regard to the neutral theory, which calls itself "non-Darwinian". The neutral theory, as a stochastic theory, takes into account only a single limiting case, namely distributions with a small population size and a huge genome. In this case, each mutant is purely accidental. The theory attempts to describe the random behaviour of each individual type. The larger the population becomes (compared with the number of positions in the sequence) and the longer the periods become over which averages are taken, the more the random behaviour is restricted by deterministic constraints. In that case, real neutrality can hardly be realized. To achieve neutrality, all mutants of two alternative distributions would have to have identical selection values which requires completely equivalent value landscapes (or they would have to compensate at least to identical eigenvalues). Conversely, small changes in the environment (temporal or local fluctuations) could, as the above example shows, be sufficient to trigger a phase change which would move the centre of gravity of the distribution in the sequence space. In strongly optimized distributions, this will have precisely the same effect as neutral drifting.

The essential difference between the classical and the new interpretation concerns the random nature of mutation. That mutations happen randomly was the dogma of classical population genetics and its extensions. In the previous chapter it was shown, however, that the appearance of a 12-error mutant of the wildtype, which finds itself in a mountainous area of the value landscape, is very much more likely (by a factor of 10^{18} in the example studied) than the appearance of any 12-error mutant which finds itself on a low value plateau. This means that the first type of mutant appears deterministically in each normal laboratory sample (for example of RNA and DNA molecules, viruses, bacteria) while the second type of mutant has hardly any chance of ever being found. With such a drastic effect (the number of the orders of magnitude can be expanded at will in models with other parameters) one must surely speak of a new quality. At the very minimum, this statement means that mutants appear in a fundamentally target-directed fashion; indeed, it means that favourable mutants have a much greater chance of appearing than (very) unfavourable or lethal ones. One must regard this as a target-directed generation of mutants. In the classical view teleonomy only arose through selection. Now we see that selection already includes some sort of advance planning of the next step in evolution. The distinction is analogous to the difference between an intelligent search for a goal and a blind try out. Jacques Monod developed his existential biological world view precisely with this alleged "blindness" of mutation in mind [17,22].

In population genetics, a mutant could only be identified when it grew to the point when it could be recognized by its phenotypical characteristics. In the age of molecular biology, one can clone and determine the sequence of individual nucleic acid molecules and genes from viruses or micro-organisms. Only with the determination of sequences is it possible to quantify the relative distance between individual mutants. It is almost a trivial insight that the frequency at which a mutant is created depends on the distance between the defective replica and the original. In the simplest case, the frequency of mutants relative to the wildtype is given by a Poissonian distribution of errors. But mutants do not arise solely from the wildtype. The favourable mutants which may be expected at relatively large distances from the wildtype will always evolve from a precursor, a mutant which is not so far away from the centre of gravity of the distribution. It is now essential to know how all the mutants — localized in the sequence space — are represented with respect to their distance-coordinates. A mutant represented by many copies will give rise to many new mutants. Thus, the occupation of sequence space is modified, not only by the Poissonian distribution of errors, but more so by the value profile, as described in the previous chapter. We can solve this problem quantitatively with the help of the eigenvalues and eigenvectors of the value matrix. It then becomes apparent that the earlier statement that mutants are generated randomly and not in a goal-oriented fashion, has no foundation any more. It is still true that singular molecular processes occur by chance; one cannot predict if and when such a process will happen, one can only say something about its probability. The statistical superposition of singular molecular processes implies, as in other areas of physics, a degree of determinism which becomes more pronounced as the population number grows. The resulting behaviour, which we call selection, can clearly be distinguished from behaviour in thermodynamical equilibrium. Unlike near-equilibrium processes, selection can be triggered by a unique singular molecular process. In this respect, we may call each historical path of evolution chance dependent. Nevertheless, chance is constrained, the overall process is value-oriented, not only by virtue of selection, but beyond that through appearance of mutants which depends upon specific occupation probabilities.

7.8 Experimental arguments

In the laboratory today, we can produce self-replicating or complementary-replicating molecular systems in vitro and these allow us to check the claims of the theory under well-defined conditions. First, the kinetics of the enzyme- and template-controlled reproduction of nucleic acids was studied experimentally and the beginnings of realistic models were developed for these processes [23]. By cloning single mutants, it was possible to study the frequency distribution within a quasi species directly and, from the frequency of back-mutation of mutants to the wildtype, it was possible to determine the error rate quantitatively [24]. And lastly, by changing the environmental conditions, selection pressure was applied which, in a series of cases, enabled the direct observation of the goal-oriented evolutionary adaptation which the change in selection pressure triggered. In all of these experiments, the results corresponded with the behaviour predicted by the theory [23,25].

Natural virus-complexes also behave in accordance with the quasi- species model; thus this concept has found its permanent place in the terminology of virologists. Initially the surprise was great when, in the case of single-stranded RNA viruses, error rates were found that were approximately one million times higher than in the case of microorganisms such as E. coli. A careful analysis revealed that viruses operate directly at the error threshold; that is, under conditions in which they can adapt themselves most quickly to any change in the environment, e.g., in the host organism's immune system. Let us consider the bacterial virus Q_β. Its genome consists of a single- stranded RNA molecule with 4200 nucleotides [25]. Its error rate was experimentally determined to be $3 \cdot 10^{-4}$ [24]. If one takes into account the relative value distribution of the mutants, one finds that the experimentally determined error rate corresponds exactly with the theoretical threshold value at which stable selection is guaranteed and evolution can be expected to occur with optimal rate. With analogous i.e., single-stranded RNA mammal viruses, similar relations are found. Influenza-A virus has an error rate of $6 \cdot 10^{-5}$, with a genome consisting of about 14.000 nucleotides [25]. On the other hand, for E. Coli cells, and especially for higher organisms, error rates smaller than 10^{-9} or 10^{-10} are found [28].

Viruses showing the greatest inclination for variation are the so called retroviruses [29], a class of viruses that carry their genetic information in an RNA molecule and transcribe it into DNA (reverse transcription or reversion, hence the name retroviruses). The information transcribed in DNA becomes integrated into the genome of the host cell. The HI-viruses, which lead to immune deficiency (AIDS), and which are single-stranded RNA viruses with a length of approximately 10.000 nucleotides, also belong to this class of retroviruses. Their error rates are of the order of 10^{-3} to 10^{-4} [30,23]. This means not only that the viruses vary in their composition from host to host, but, also, that even in a single host cell a broadly scattered spectrum of mutants (a typical quasi-species distribution) is already present. An immediate consequence of this property is that it is very difficult to eliminate this virus. It evades the immune response by continuous modification, in a manner similar to that already observed for Influenza-A virus [31], but with even greater efficiency. Moreover, the control centre of the immune system itself appears to be one of the targets of this virus.

In the context of our considerations about the theory of evolution, we are directly confronted with one of the great challenges of the present. For the application of these newly found insights in practical situations, the future has already started. Thus, at the Max-Planck-Institute in Göttingen, we are currently developing a series of evolution machines with the help of which we try to realize evolutionary adaptation under optimal conditions. For this purpose, single mutants have to be cloned and multiplied under variable error conditions. One of the intended applications concerns the identification of the unchangeable, invariant positions on virus genomes. Knowledge of these positions would make it possible to stimulate an immune response much more directly [17,32–35].

However, our interest is not directed solely at the future. Modern methods for comparing sequences enable a more precise analysis of the past and of our origin, the origin of life per se. For this, it is essential to compare related sequences from different stages of evolutionary development with one another and to locate them correctly in the multi-dimensional sequence space. This new method allows the dating of evolutionary events that took place long before the first cell branches were established [15,16,17]. The genetic code, the basis of information processing in all the different stages of life, emerged about 3.5 to 4 billion years ago according to these studies. Our planet is about 4.7 billion years old. The qualitative statement, that there was not enough time for the development of life on our planet, is thus, for the first time, refuted by quantitative arguments.

7.9 From the specific to the general

In the end we are, of course, more interested in the relevance of these new insights than of their specific applications. The quasi-species model as such is of a very specific nature and, consequently, limited in its quantitative interpretation. This should be considered an advantage rather than a disadvantage. It also allows concrete general statements to be formulated in the precise vocabulary of physics. We are dealing with new parameters: quantity of information, error rate, value landscape; these have their direct counterpart in physical parameters like temperature (melting point), entropy, potential surface, etc. Beyond that, the derived relations show a structure familiar to the physicist. Generation of information, for example, follows the structure of the (two-dimensional) Ising-model of ferromagnetism. On the other hand, the quantitative applications of the model are limited exactly to those systems which fulfil the assumptions of the model precisely, that is, limited to replicative systems of the size of a gene or a virus. Already the application to primitive autonomous micro-organisms necessitates an expansion of the model: genes must be integrated functionally through cyclical (functional) couplings between the cycles of replication, cycles in which the function of the gene is coupled back to its generation and preservation. We call this superposition of cycles "hypercycles". These, in turn, could only optimize their functions by being brought together spatially through compartmentalization. It was still a long way to the autonomous cell, a unit which, by virtue of the properties of reproduction of nucleic acids, multiplies itself synchronously as a whole. Further phase transitions follow thereafter. Cells start to communicate with each other, to differentiate and to build up organs. Life becomes visible as a form. An organ for communication grows out of the interaction between single cells. Individuals equipped with such an organ start to exchange and store information. Eventually, the cultural memory of a social entity of individuals grows out of the individual memories.

In this context, we must warn that a model developed from specific assumptions cannot be applied readily in situations where more complex assumptions are needed. But this warning can immediately be weakened: the higher levels of self-organization are similar to one another in essential aspects; they differ only in the details of their organization. Surprisingly, mathematical models that have been developed to describe cell differentiation or even certain functions of nerve networks, are structurally related to each other and to the above model of evolution. The model developed here is precisely typical for the transition from a material structure to immaterial information. Information has phase properties. The information contents of different phases is separated in a manner analogous to the spatially coherent material phases. Information generation takes place through phase changes in which sudden developments occur which must be delineated and saved. This terminology, borrowed from physics, goes much further than the metaphorical, as is shown by the concrete model.

The theory of the quasi-species model also has the "if – then" character typical of all physical theories. Basically it does not say anything about the real origin of life, about the historical genesis of biological information. It does indicate what one should take into account when one wants to understand this process, which for a long time was considered "ultra vires" [36] because of the complexity of the problems that needed to be overcome. It also indicates that many of the detailed interpretations of biological reality based on Darwinian theory need to be altered. The theory merely points to some aspects of reality without representing it in detail. Only in this sense may we interpret the application of the idea of phase changes to such complicated phenomena as the creation of information in the human brain, in society and in culture.

References

[1] R.W. Millar, J. Am. Chem. Soc. **50** (1928) 1875.

[2] H. Haken, *Erfolgsgeheimnisse der Natur*, Deutsche Verlags Anstalt, Stuttgart, 1986.

[3] H. Haken, *Synergetics. An Introduction*, Springer, Heidelberg, 1983.

[4] S. Grossmann, Phys. Blätter **39** (1983) 139; H.G. Schuster, *Deterministic Chaos — An Introduction*, Physik Verlag, Weinheim, 1984.

[5] H.O. Peitgen and P. Richter, in *The Beauty of Fractals*, Springer, Berlin, 1986.

[6] C. Darwin, *The Origin of Species*, 1859, New edition, Crowell–Collier Publishing Comp., Toronto, 1962.

[7] I. Rechenberg, *Evolutionsstrategie*, Problemata frommann–holzboog, Stuttgart–Bad Canstatt, 1973.

[8] M. Eigen and R. Winkler-Oswatitsch, Naturwissenschaften **68** (1981) 217, 282; M. Eigen, B. Lindemann, R. Winkler-Oswatitsch and C.R. Clarke, Proc. Natl. Acad. Sci. U.S. **82** (1985) 2437.

[9] R.W. Hamming, *Coding and Information Theory*, Prentice Hall Inc., Englewood Cliffs, 1980.

[10] M. Eigen, Ber. der Bunsenges. phys. Chem. **89** (1985) 658.

[11] P. Schuster and K. Sigmund, Ber. der Bunsenges. phys. Chem. **89** (1985) 668.

[12] M. Eigen, Chemica Scripta **26B** (1986) 13.

[13] P. Schuster, Chemica Scripta **26B** (1986) 27.

[14] M. Eigen, J. McCaskill and P. Schuster, J. Phys. Chem. **92** (1988) 6881.

[15] R. Winkler-Oswatitsch, M. Eigen and A. Dress, Chemica Scripta **26B** (1986) 59.

[16] A summary has been published in Proc. Natl. Acad. Sci. U.S. **85** (1988) 5913 and Science **244** (1988) 673.

[17] M. Eigen, *Stufen zum Leben*, Piper, München, 1987.

[18] B.B. Mandelbrot, *The Fractal Geometry of Nature*, Freeman, New York, 1983.

[19] J. Swetina and P. Schuster, Biophys. Chem. **16** (1982) 329.

[20] M. Kimura, *The Neutral Theory of Molecular Evolution*, Cambridge University Press, Cambridge, 1983.

[21] J.L. King and T.H. Jukes, Science **164** (1969) 788.

[22] J. Monod, *Le hazard et la nécessité*, Edition du Seuil, Paris. German edition: *Zufall und Notwendigkeit*, Piper, München, 1971.

[23] C.K. Biebricher and M. Eigen, *RNA-Genetics*, eds. E.Domingo, P. Ahlquist and J.J. Holland, CRC Press, Baton Rouge, 1988.

[24] E. Domingo, D. Sabo, T. Taniguchi and Ch. Weissmann, Cell **13** (1978) 735.

[25] C.K. Biebricher, *Evolutionary Biology*, Vol. 16, eds. M.K. Hecht, B. Wallace and G.T. Prance, Plenum Press, New York, 1983.

[26] P. Mekler, Dissertation, Universität Zürich, 1981.

[27] J.D. Parvin, A. Moscona, W.T. Pan, J.M. Leider and P. Palese, Journal of Virology **59** (1986) 377–383.

[28] J. Maynard-Smith, *Evolution and the Theory of Games*, Cambridge University Press, Cambridge, 1982.

[29] D. Baltimore and H.M. Temin, *Les Prix Nobel*, Stockholm, 1975.

[30] R.C. Gallo, Scientific American **256** (1987) 38.

[31] P. Palese, *Evolutionary Processes and Theory*, eds. S. Karlin and E. Nevo, Academic Press, New York, 1985.

[32] M. Eigen, *Evolution of Catalytic Function*, Cold Spring Harbor Series, 1987.

[33] Y. Husimi and H.C. Keweloh, Rev. Sci. Instrum. **58** (1987) 1109; A. Schwienhorst, Diplomarbeit, Göttingen/Münster, 1987.

[34] H. Otten, Dissertation, Göttingen/Braunschweig, 1987.

[35] G. Bauer, Dissertation, Göttingen/Braunschweig, 1987.

[36] C. Darwin, in a letter to N. Wallich, cited in [17], p. 56.

Between Science and Technology
A. Sarlemijn and P. Kroes (Editors)

Chapter 8

From laser physics to synergetics and pattern recognition

H. Haken

Institute for Theoretical Physics and Synergetics
Stuttgart

8.1 Introduction

The study of cooperative effects is certainly one of the central topics of theoretical physics in this century. One important phenomenon of this kind is superconductivity, which by the recent discovery of high T_c superconductivity by Bednorz and Müller is finding a largely revived interest. When I was a student at Erlangen, we had to give seminar talks on this field at a time when an adequate microscopic theory was still lacking. However, as it turned out, phenomenological theories were of great importance and among the prominent ones were those by Casimir and Gorter [1]. So at that time I came across for the first time with Prof. Casimir's name. In my talk I shall address the topic of cooperative phenomena also, but in systems driven far from thermal equilibrium. In this context the laser has played an eminent role. It has proved to be a practically inexhaustible source of inspiration and new insights. In addition it turned out to be a very useful model system against which various concepts of statistical physics and thermodynamics can be checked. Here I hardly need to stress the enormous technical applications of the laser, which for instance in CD-players is bringing music into many homes.

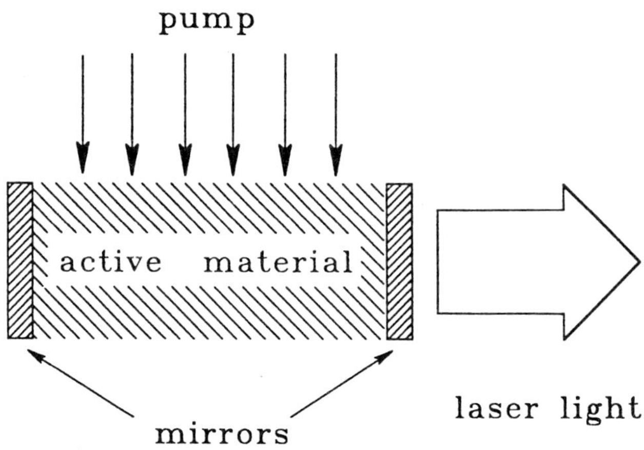

Figure 8.1: *Typical experimental set-up of a laser.*

My paper will be organized as follows: first I shall present some important aspects of laser theory, then I will show how some of the concepts can be generalized to much more general systems, and finally I wish to show how these concepts can be applied to construct parallel computers for pattern recognition. I shall present explicit examples of pattern recognition by the algorithm I developed on the basis of synergetics.

8.2 Some basic concepts of laser theory

The experimental setup of a laser is well-known; it consists of a laser active material which may be, for instance, impurity atoms like chromium ions, embedded in a crystal in an aluminium oxide forming ruby [2]. The laser has the shape of a rod, at the end-faces of which two mirrors are mounted (fig. 8.1). They serve to select light waves running in axial direction. Because these waves are reflected more often than any other waves and thus have a longer life-time, they interact more intensely with the laser active atoms. In addition a selection with respect to wave length is done because only standing waves with $n\lambda/2 = L$ fit inbetween the two mirrors, where n is an integer, λ the wavelength and L the distance between the mirrors. The laser atoms are pumped from the outside, e.g. by an electrical current in a gas discharge or by shining light from the outside onto these atoms. In order to bring out some of the most essential features, we shall be concerned with a single-mode laser in which only one type of wave is supported by the laser active atoms. In order to derive a satisfactory theory, we shall take into account both the nature of the quantum mechanical "motion" of the electrons in the atoms as well as the quantum nature of light [3,4]. In order to take into account the latter, we decompose the electric field strength, which depends on space x and time t according to

$$E(x,t) = (b(t) + b^+(t)) \sin kx \, . \tag{8.2.1}$$

The space-dependent function $\sin kx$ is determined by the geometry, whereas the amplitude in front of this function is split into the positive and negative frequency parts, respectively.

$$b \propto e^{-i\omega t}, \; b^+ \propto e^{i\omega t} \, . \tag{8.2.2}$$

In a quantum mechanical treatment b and b^+ are the annihilation and creation operators of a photon, respectively, and obey the well-known Bose–Einstein commutation relations for Bose-operators. As is well-known from quantum mechanics, the Hamiltonian of the field acquires the form

$$H_{field} = \hbar\omega b^+ b \tag{8.2.3}$$

where ω is the frequency of the field mode. The essential features of laser action can be brought out if we consider the laser atoms as having only two levels, one ground-state level and one excited level. It has turned out to be advantageous to employ the methods of second quantization so that we introduce the operators a_1, a_2 describing the annihilation of an electron in the states 1 and 2, respectively, and the corresponding creation operators a_1^+, a_2^+ [3,5]. Denoting further the energies corresponding to the two states 1 and 2 by W_1, W_2, respectively, the Hamiltonian of the atom under consideration acquires the form

$$H_{atom} = W_1 a_1^+ a_1 + W_2 a_2^+ a_2 \, . \tag{8.2.4}$$

In laser theory it has turned out to be further advantageous to introduce instead of the creation and annihilation operators a, a^+ the quantized transition amplitudes which are defined by

$$\alpha^+ = a_2^+ a_1 \tag{8.2.5}$$

and

$$\alpha = a_1^+ a_2 \, . \tag{8.2.6}$$

As we shall see below, they can be interpreted directly as operators of the atomic dipole moment and thus be visualized in a very simple fashion. It is now rather simple to rewrite the Hamiltonian of a single atom in the form

$$H_{atom} = W\alpha^+\alpha \tag{8.2.7}$$

where

$$W = W_2 - W_1 \tag{8.2.8}$$

is the energy difference between the upper and lower atomic level. In general, laser light will be generated by the cooperation of many laser active atoms. We shall distinguish between them by an index μ so that (8.2.7) has to be replaced by a sum over the individual contributions of these atoms

$$H_{atoms} = \sum_{\mu} W \alpha_{\mu}^{+} \alpha_{\mu} \,. \tag{8.2.9}$$

Finally, we have to consider the interaction between the atoms and the field mode (1). Leaving aside virtual transitions which do not play a role in usual laser action the interaction Hamiltonian can be described in the following way: when an electron in the upper state 2 is annihilated and makes a transition to its lower state, whereby it is generated in that state by the operator a_1^+, a photon is created which is described by the creation operator b^+. Taking into account the reverse process of the annihilation of a photon also, we are led to the Hamiltonian

$$H_{int} = \hbar g (a_2^+ a_1 b + a_1^+ a_2 b^+) \tag{8.2.10}$$

which can be derived from first principles of quantum electro-dynamics, also. g is a coupling constant which describes essentially the strength of the dipole moment, whereas $\hbar$ is Planck's constant divided by 2π. Introducing the operators α and α^+ and summing up over all the atoms, we are led to the interaction-Hamiltonian

$$H_{int} = \hbar \sum_{\mu} g (\alpha_{\mu} b^+ + \alpha_{\mu}^+ b) \,. \tag{8.2.11}$$

The Hamiltonian of the total system then reads

$$H_0 = H_{field} + H_{atoms} + H_{int} \,. \tag{8.2.12}$$

At first sight it may appear that the Hamiltonian (8.2.12) is sufficient to describe the laser system and, in fact, a number of authors have tried to base laser theory on this Hamiltonian solving the corresponding Schrödinger equation. However, an important aspect is left out. The laser is an open system into which energy is fed all the time, e.g. by the pump light, in which irreversible processes take place due to the coupling of the atoms to their surrounding, and which emits the laser light. To describe this interaction of the atoms and the field with the surrounding, the dynamics of so-called reservoirs or heat baths must be encountered.

Therefore the Hamiltonian (8.2.12) must be supplemented by a Hamiltonian, $H_B + H_{B-L}$, describing the interaction with these reservoirs.

$$H = H_0 + H_B + H_{B-L} \,. \tag{8.2.13}$$

Again it may appear that now the task is to solve the Schrödinger equation belonging to (8.2.13). However, (8.2.13) contains a great many variables of the heat baths which we are not interested in. Thus the task arises to eliminate the heat bath variables in an adequate fashion. Their effect may be guessed from the fundamental work of Langevin, Einstein and others. Namely, when a molecule is immersed in a liquid or gas which are described by many degrees of freedom, we may explicitly or implicitly eliminate these variables and find, eventually, an equation for the molecule coordinates alone where new terms in the form of damping and fluctuating forces arise. Over the past decades a sizable literature has arisen how to eliminate bath variables to which Bose or Fermi fields are coupled and various elimination schemes were developed based on Heisenberg equations of motion for quantum mechanical operators or on the density matrix. I shall not dwell here on any of these details because they are of no basic concern to my paper here. As is well-known the Heisenberg equations for any operator are given by

$$\dot{\Omega} = \frac{i}{\hbar}[H, \Omega] \,. \tag{8.2.14}$$

where the bracket denotes the commutator. When this prescription is applied to the field operator b and the atomic dipole operator α_μ and the heat bath operators have been eliminated, the equations

$$\dot{b} = (-i\omega - \kappa)b + i\sum_\mu g\alpha_\mu + F \tag{8.2.15}$$

and

$$\dot{\alpha}_\mu = (-i\omega - \gamma)\alpha_\mu - igbd_\mu + \Gamma_\mu \tag{8.2.16}$$

are obtained. Let us discuss the individual terms of these equations in detail. According to equation (8.2.15) the temporal change of b is determined by the following terms: $-i\omega b$ describes the free oscillation of the field amplitude. $-\kappa b$ represents the damping of the field amplitude due to losses. The sum over μ describes the driving of the field by the oscillating dipole moments where the slowly varying amplitude approximation is used. Finally, F represents the fluctuating forces. According to the fluctuation-dissipation theorem, such fluctuating forces appear whenever damping occurs. This relationship is not only true in classical physics but extends also to quantum mechanics in a specific fashion. One should note that b, α_μ and also the fluctuating force F are operators though we shall not enter a discussion of the quantum mechanical properties of F.

Let us now consider equation (8.2.16). The first term on the right-hand side describes the free oscillation of the dipole moment, the term proportional to γ describes its damping which can be related to the optical line width. The term proportional to g describes the driving of the amplitude of the atomic dipole which is proportional to the field operator b and in addition to the inversion d_μ where d_μ is the difference of the occupation number of the upper and lower levels, respectively. Γ_μ are again fluctuating forces which are related to γ. Because of the appearance of the operator d_μ a third type of equations must be derived for that operator which again by use of (8.2.14) acquire the form

$$\dot{d}_\mu = \gamma_{\|}(d_0 - d_\mu) + 2ig(\alpha_\mu^+ b - \alpha_\mu b^+) + F_{\mu o} \,. \tag{8.2.17}$$

The first bracket describes the relaxation of the inversion towards the equilibrium inversion d_0 which is a c number and which is described by incoherent pump and relaxation processes. The middle bracket describes the change of the inversion because of the coupling of the field to atomic dipole moment whereas the last term represents fluctuating forces. The equations (8.2.15) - (8.2.17) are the basic quantum mechanical equations of the single mode laser. In my early work [6] I showed how they can be solved directly in regions below and above the so-called laser threshold. Because b, α_μ and d_μ are operators, their handling requires some care. However, as it has turned out, the equations (8.2.15) - (8.2.17) can be translated into classical equations where b, α_μ and d_μ are then classical quantities having the meanings already used above. Though there is an extended literature on the translation method from quantum mechanics to classical mechanics, which is called quantum classical correspondence, space does not allow me to present this quite interesting translation formalism [7]. At any rate we shall be on safe ground to interpret the equations (8.2.15) - (8.2.17) in the following as classical equations.

At first sight it seems rather hopeless to solve these equations even on a large computer for two reasons: first of all they are very numerous, there are e.g. 10^{18} atoms in a laser, and they are nonlinear as expressed by the terms bd_μ, $\alpha^+ b$ and $\alpha_\mu b^+$. However we shall see in a minute, and this is the first insight provided by laser theory, how this large set of equations can be reduced considerably. To this end we first eliminate the time-dependence prescribed by the frequency ω by means of the hypothesis

$$b = B(t)e^{-i\omega t} \tag{8.2.18}$$

and

$$\alpha_\mu = A_\mu(t)e^{-i\omega t}. \tag{8.2.19}$$

The equations (8.2.15) and (8.2.16) then acquire the form

$$\dot{B} = -\kappa B + i\sum_\mu gA_\mu + \tilde{F}_\mu \tag{8.2.20}$$

and

$$\dot{A}_\mu = -\gamma A_\mu - i\sum_\mu gBd_\mu + \tilde{\Gamma}_\mu \tag{8.2.21}$$

where the fluctuating forces differ from those in (8.2.15) and (8.2.16) by a factor $e^{i\omega t}$. We now employ the fact that in most lasers the atomic damping constant γ is much bigger than that of the field mode b:

$$\gamma \gg \kappa\,. \tag{8.2.22}$$

Though the following procedure may be cast in a rigorous form, a few hand-waving arguments may suffice here. Let us consider the first term on the right-hand side of equation (8.2.20) which describes damping. Because κ is much smaller than γ, the decay of B occurs very slowly. Now in equation (8.2.21) B occurs as a driving force on the right-hand side. Neglecting a short transient period we may thus state that the amplitude A_μ is driven by B and decays at the same rate as B does. Thus we immediately find the inequality

$$|\dot{A}_\mu| \ll \gamma|A_\mu| \tag{8.2.23}$$

verified because of (8.2.22). Because the left-hand side is much smaller than the right-hand side, we may ignore the time derivative of A_μ in equation (8.2.21) which allows us to express A_μ directly by B and d_μ for which a similar argument holds. The possibility to express A_μ is here based on the argument of adiabatic following (A_μ follows adiabatically B, a concept introduced at an early stage in particular by Ehrenfest). As I could show within the past two decades, this principle allows for a considerable extension, namely it can be shown that A_μ can be expressed explicitly and exactly by B and d_μ. This mathematical procedure is called the slaving principle and it possesses some special cases known in mathematics as central manifold theorem. This slaving principle is a very pleasant consequence, namely it allows us to express the very numerous degrees of freedom represented by A_μ by a single degree of freedom B. We

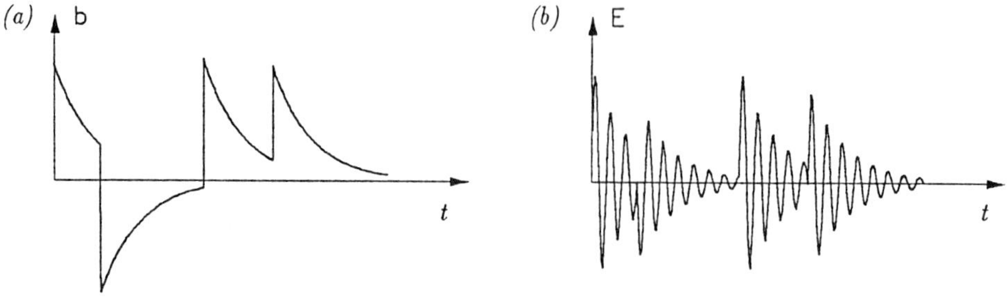

Figure 8.2: *(a) The field envelope versus time. (b) The field strength, i.e. the envelope multiplied with* $\sin \omega t$ *versus time.*

mention that similar arguments hold for the inversion d_μ which eventually can be expressed by B and A_μ because of

$$\gamma \gg \gamma_{\|} \gg \kappa . \tag{8.2.24}$$

In this way all the atomic variables can be eliminated and we end up with an equation for the field amplitude B alone. The equation

$$\dot{B} = (G - \kappa)B - CBB^{+}B + F_{tot} \tag{8.2.25}$$

is the basic equation which we wish to discuss now. It has three features we shall come across again and again in the field of synergetics. The behaviour of the solution will depend critically on the sign of $G-\kappa$. $G-\kappa$ thus serves as a "control parameter" which controls the behaviour of the system. The second term on the right-hand side of equation (8.2.25) is nonlinear, again a typical feature of synergetic systems. And finally stochastic or fluctuating forces F_{tot} occur. This equation cannot be solved in closed form and therefore I wish to discuss some approximations first. If the laser acts as a usual lamp, we may suppose that the field amplitude is a small quantity so that the nonlinear term can be neglected. If the control parameter $G-\kappa$ is negative, the following behaviour results: the fluctuating forces F act as a sequence of kicks given to the field amplitude B which after each kick relaxes towards zero. Thus we find the behaviour of the field amplitude as depicted in fig. 8.2a and if we take into account the rapid oscillations of b the behaviour in fig. 8.2b. Clearly when we pump the laser more strongly, the gain factor G increases so that the difference $G-\kappa$ becomes smaller what its absolute value is concerned. In other words, the individual wave-tracks decay more and more slowly. As may be shown in optics, the longer decay of the field amplitude results in a smaller line width. Thus, we arrive at the result that with increasing pump power and correspondingly

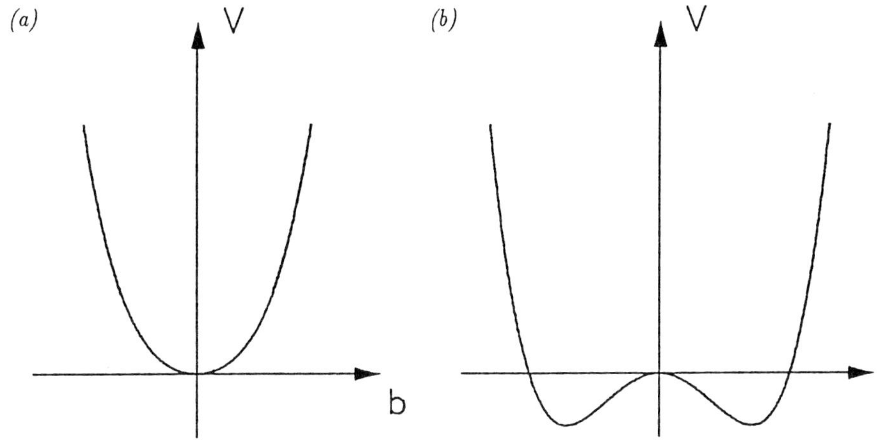

Figure 8.3: *(a) Visualization of the potential as a function of the field amplitude for a negative net gain. (b) The same as in (a), but for positive net gain.*

with increasing output power, the line width decreases according to the formula

$$\Delta\nu = \frac{Const}{P}. \tag{8.2.26}$$

which is the famous Townes' formula for laser and maser action where P is the output power of the laser. But what happens when the output power becomes so large that the nonlinear term in (8.2.25) can no more be neglected? In order to discuss this case I resorted to an analogy with mechanics, namely when we add the term $m\ddot{B}$, i.e. an acceleration term of a particle with mass m to the l.h.s. of (8.2.25), this equation can be interpreted as the equation of motion of a particle with a damping force given by $\dot{B}$ and being subject to a deterministic force described by the two terms of the r.h.s. of equation (8.2.25) and a stochastic force described by F_{tot}. When the damping is relatively large, we may again safely neglect the acceleration term $m\ddot{B}$ and return to equation (8.2.25). But now we have a simple interpretation of this equation at hand, namely that of the motion of an overdamped particle in a potential field and subject to stochastic kicks. This potential field is plotted for negative and positive values of $G - \kappa$ in fig. 8.3. In fig. 8.3a, i.e. for negative $G - \kappa$, we arrive at a situation already described above. The fluctuating forces act like a team of soccer players who kick the ball at random. After each kick it rolls down to the bottom of the valley and again a figure of the kind 8.2 results. When $G - \kappa$ becomes smaller, the slope of the mountains becomes flatter and flatter, the ball rolls down more slowly, a phenomenon known in phase transitions as

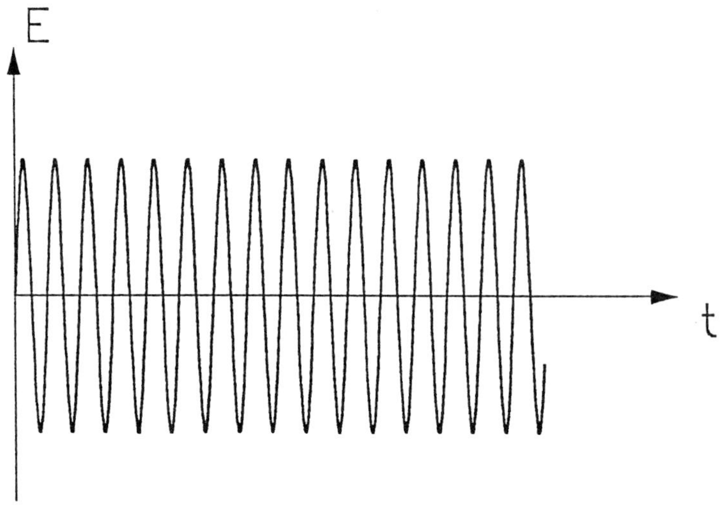

Figure 8.4: *The coherent laser wave (field strength versus time).*

critical slowing down. But from the fig. 8.3b it is evident that we must expect a qualitatively entirely different behaviour of the field amplitude when $G - \kappa$ becomes positive. In this case we expect a stable field amplitude, and when we consider $b = Be^{-i\omega t}$, an oscillation with the frequency ω and a stable amplitude. Superimposed on this oscillation are small amplitude and phase fluctuations caused by the fluctuating forces. Thus the behaviour of laser light should be quite different from that below threshold. The formerly incoherent uncorrelated wave-tracks are now replaced by a single amplitude-stabilized field mode (fig. 8.4). When I discussed my predictions at that time in 1964 with some laser experts, they told me that this case cannot happen because I am dealing here with an "unphysical region". They told me that when $G - \kappa$ increases from negative values to 0, the field amplitude will increase more and more and will reach an asymptotic value described by a potential curve which is becoming flatter and flatter. However, I was so much convinced of my results that I published them in spite of the critics of the experts [6]. A short time later my predictions were fully substantiated experimentally by measurements of Armstrong and Smith [8] on semi-conductor lasers. Since that time I came to the conclusion that even experts may fail and one should trust only one's own results. The predictions I made on this statistical behaviour of laser light were made in the form of correlation functions on phase and amplitude or, correspondingly, on intensity fluctuations. But when there are moments or correlation functions, there must be the corresponding distribution function. In their derivation the decisive step was done by my former co-worker Risken [9] who interpreted (8.2.25) again as a classical equation and established the Fokker–Planck equation corresponding to the Langevin type equation (8.2.25). Here were quickly found that the distribution function for laser light is given by

$$f(B) = N\exp(\frac{G-\kappa}{2}|B|^2 - \frac{C}{2}|B|^4)\,. \tag{8.2.27}$$

This function has been checked experimentally by a number of authors, notably by Pike [10] and Arecchi [11] and their co-workers, and excellent agreement has been found. In other words, the quantum mechanical and statistical theory of laser light has been checked in great detail and with utmost precision.

8.3 The strange behaviour of entropy and information

Let me now make a big jump from the early days of laser theory to more recent developments. As was stated before the laser is an open system being pumped from the outside. It is thus driven far from thermal equilibrium which allows us to check a number of concepts of thermodynamics or statistical physics against this model system. Let me pick out only one specific example which I believe is rather dramatic. It has been stated by some well-known authors that when a system far from thermal equilibrium goes from its disordered to its ordered state (which is the case with the laser) then the change of entropy

$$dS < 0\,. \tag{8.3.1}$$

In this author's opinion the use of thermodynamic entropy in systems far from thermal equilibrium is anyway dubious because according to its definition we need the absolute temperature T in the standard equation

$$dS = \frac{dQ_{rev}}{T} \tag{8.3.2}$$

and we know quite well, for instance from the laser, that such a system does not possess any unique temperature. Thus, to be on safe grounds, one has to use the entropy in the statistical sense which may be given by the Boltzmann formula

$$S = k\ln W \tag{8.3.3}$$

or in a related fashion in the form of information according to Shannon [12]

$$i = -\sum_j p_j \ln p_j\,. \tag{8.3.4}$$

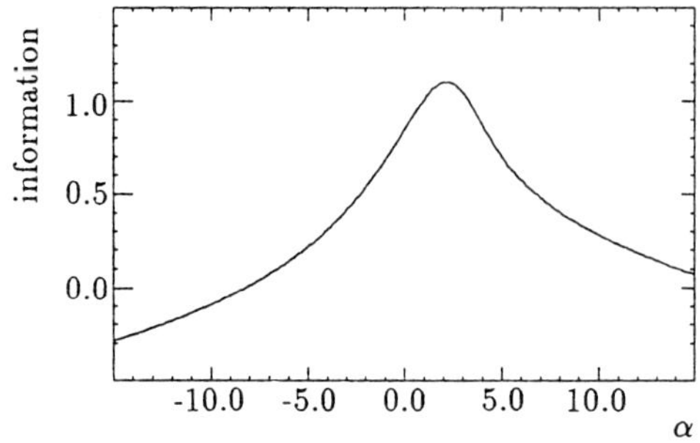

Figure 8.5: *Behaviour of the information versus the control parameter α. At $\alpha = 0$ the instability occurs.*

Here p_j is the probability distribution of the system over states labelled by the index j. In order to express (8.3.4) in bits we must replace ln by $\log_2$

$$\ln \rightarrow \log_2 \tag{8.3.5}$$

which is of minor concern here, however. Let us first discuss i when we utilize the distribution function (8.2.27). In order to make the proper identification, we have to replace the sum over j by an integral over B, and we have to identify the index j with the amplitude B. When B is a real quantity, which happens if B belongs to a laser phase-locked to an external source, i has the behaviour shown in fig. 8.5 as a function of the control parameter $G - \kappa$. $G - \kappa = 0$ corresponds to the laser threshold. As we see well below and well above laser threshold, the information differs quite appreciably and *has increased*. The enhanced value of i in the transition region results from critical fluctuations. When, on the other hand, B is a complex quantity as it happens in a usual laser, then it turns out that the behaviour of fig. 8.6 results. Again the information or, equivalently, the entropy increases when the laser goes from its disordered into its ordered state in contradiction to the intuitive predictions done by people working in thermodynamics. One may complain that here we studied only the statistical properties of laser light and not of the total laser. Therefore we studied also the behaviour of the information of the total laser, i.e. laser light plus laser active atoms. Because of the slaving principle the probability distribution p can now be written in the form

$$p_{ij} = f(B)\, P(\alpha_\mu | B) \tag{8.3.6}$$

where the first function f is identical with (8.2.27) whereas P represents the conditional probability that the dipole moments α_μ acquire specific values provided the value of B is

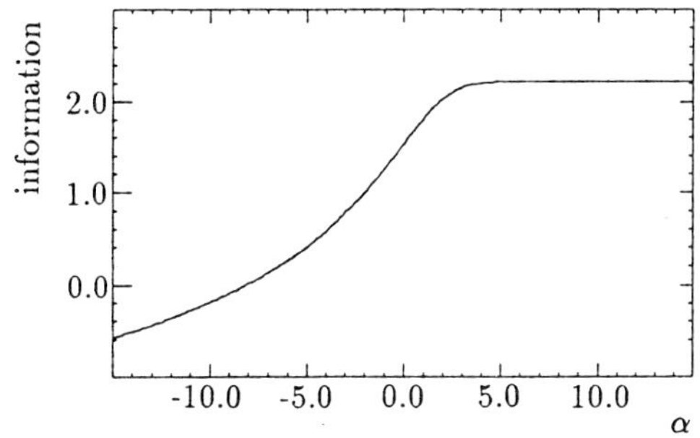

Figure 8.6: *The information as a function of the control parameter for the laser field. Note the pronounced increase of the information at the laser threshold at $\alpha = 0$ even in spite of the fact that the laser acquires an ordered state.*

prescribed. As a consequence of (8.3.6) the information (8.3.4) can be split into two parts, one referring to the light field alone and another part describing the information of the atoms where an average is taken over the field distribution function. As a detailed evaluation of

$$i = i_{field} + i_{atoms} \tag{8.3.7}$$

in the laser case shows, the contribution of the atoms to the information is orders of magnitude smaller than that of the field mode [14,15]. Thus, not only the dynamical behaviour of the laser close to threshold but also its information is governed by the field alone. And again we find that the equation

$$dS > 0 \tag{8.3.8}$$

holds if the laser goes from its disordered into its ordered state.

8.4 Some basic concepts of synergetics

The laser is a prototype of a system of synergetics [16,17]. It is composed of many individual parts (the atoms). It is a system far from thermal equilibrium, it is open, and it shows a qualitative transition between disorder and order. The ordered structure produced is the coherent laser light. Actually, there are many systems in physics, chemistry, biology, and other sciences in which many systems cooperate so that they produce spatial, temporal or functional structures at a macroscopic level. Examples are provided by structure formation in fluids, e.g. the famous Bénard instability, in chemistry the formations of rings, spirals, and oscillations by chemical reactions, in biology morphogenesis, coordination of motions of limbs, etc. As it has turned out, the basic concepts of laser physics, which are outlined before, are again of central importance here. These concepts were control parameter, instability, emergence of order parameters, and the slaving principle. There is, however, one distinction between laser physics and most of these other systems. Namely, while in laser physics there is a clearcut distinction between decay-times for the field mode, κ, and the atoms $\gamma, \gamma_{\|}$, in most other systems such a clear distinction is not there from the beginning. For instance when a fluid becomes unstable when it is heated from below, all the individual molecules have the same relaxation time and it is not obvious how a specific damping constant, which is much smaller than the other ones, can occur. In order to clarify this point, let us now give an outline of the general approach. We describe the state of a system by a state vector which may depend with its components on space and time and whose individual components may be e.g. density, temperature and velocity in a fluid, densities of chemical reactants, and so on. Denoting the state vector by q, the general form of the equation is

$$\dot{q} = N(q, \alpha) + F(t) \tag{8.4.1}$$

where N is a nonlinear function depending on the state vector q and a control parameter value α_0. This solution will be denoted by q_0 and may be space-dependent and in addition it may be constant, periodic or quasiperiodic in time. In the following we shall consider, however, the case that q_0 is time-independent (for generalizations cf. the book [18]). We now change the control parameter α_0 towards a new value α continuously and study whether the old solution remains stable or not:

$$\alpha \, : \, q = q_0 + w \, . \tag{8.4.2}$$

To make the instability analysis we use standard procedures by inserting (8.4.2) into (8.4.1) and linearizing (8.4.1) with respect to w. The resulting solutions can be written in the form

$$w = e^{\lambda t} \, v(\mathbf{x}) \tag{8.4.3}$$

where we shall denote those configurations where

$$\mathrm{Re}\, \lambda \geq 0 \tag{8.4.4}$$

as unstable modes, those where

$$\mathrm{Re}\, \lambda < 0 \tag{8.4.5}$$

as stable modes. We now use the spatial functions v as a new frame of reference. In order to solve (8.4.1) we insert the hypothesis

$$q = q_0 + \sum_{\mu} \xi_\mu(t) v_\mu(\mathbf{x}) \tag{8.4.6}$$

into these equations which can be transformed into equations for the mode amplitudes ξ alone:

$$\dot{\xi}_\mu = \lambda_\mu \xi_\mu + N_\mu(\xi_\mu) + F_\mu \,. \tag{8.4.7}$$

These equations are fully equivalent to the original equations (8.4.1) and at first sight it seems that there is no advantage in using these equations over that of using (8.4.1). However, we may now distinguish the equations according to

$$\mathrm{Re}\,\lambda_\mu \geq 0, \tag{8.4.8}$$

$$\mathrm{Re}\,\lambda_\mu < 0. \tag{8.4.9}$$

When the system is close to threshold, the λ_μ obeying (8.4.8) are quantities small compared to the absolute values of (8.4.9). In such a case the slaving principle holds which allows us to express the amplitudes of the slaved modes ξ_s in a rigorous and explicit fashion by the mode amplitudes ξ_u of the unstable modes so that we may eliminate ξ_s according to

$$\xi_s = f_s(\xi_u, t) \,. \tag{8.4.10}$$

By means of this slaving principle it becomes possible to reduce the enormous number of degrees of freedom to the dynamics of very few degrees of freedom only which obey equations of the form

$$\dot{\xi}_u = \lambda_u \xi_u + \tilde{N}_u(\xi_1, \ldots, \xi_M, t) + F_u(t) \tag{8.4.11}$$

which result from inserting (8.4.10) into (8.4.7) for (8.4.8). Quite a number of systems obey equations of the form we encountered in laser physics, namely

$$\dot{\xi}_u = \lambda_u \xi_u - \xi_u^3 + F_u, \quad u = 1 \tag{8.4.12}$$

which can then be discussed in the same terms as before. But also more complicated equations of the type (8.4.11), where $u = 1, \ldots, M$, have been discussed and allowed us to identify a number of classes of behaviour of seemingly different systems. A number of results are exhibited in the fig. 8.7–8.10 showing the power of the general approach. In particular it turns out that order parameter equations of the form (8.4.11) can give rise to complicated

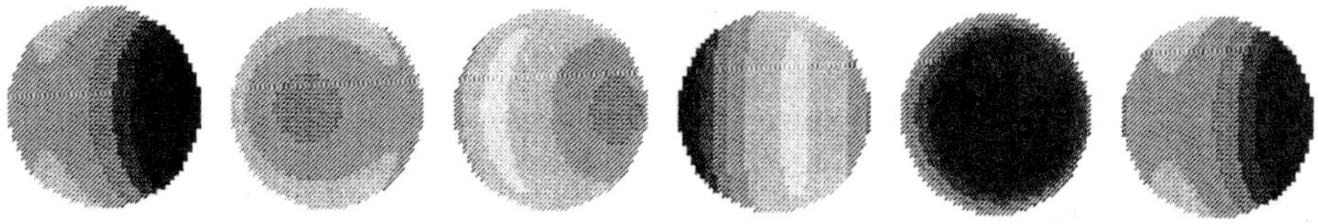

Figure 8.7: *Model of atmospheric motion. A layer of liquid on a sphere is subjected to gravitation and a heating at its inner surface and a cooling at its outer surface. In spite of the homogeneous excitation, a rotating wave of motion with a temperature field is generated. The different shapes indicate the values of the temperature (after Friedrich and Haken [19]).*

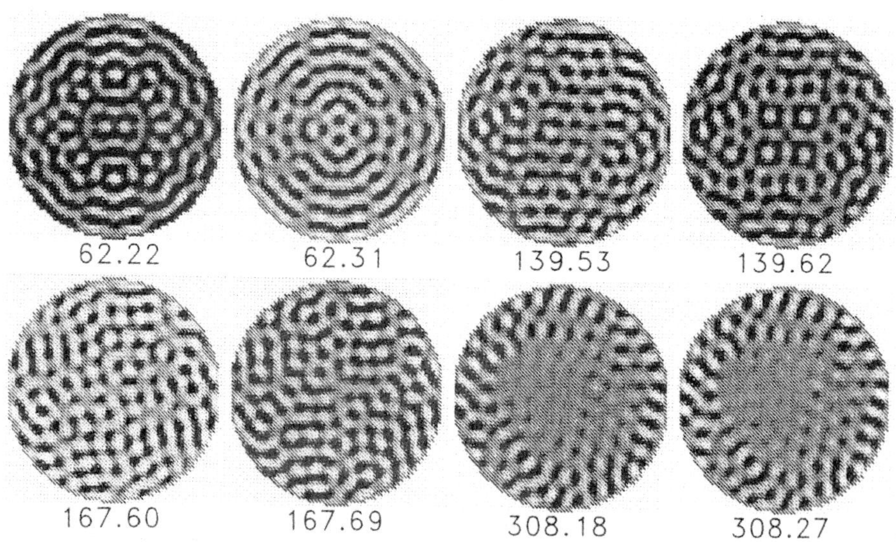

Figure 8.8: *Example of the calculation [20] of the formation of a pattern in a binary fluid mixture which is heated from below (the numbers indicate times measured in units of vertical diffusion time).*

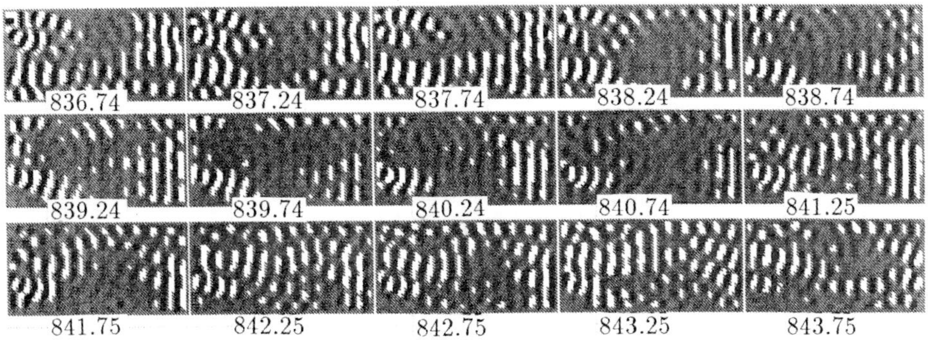

Figure 8.9: *Pattern formation in a binary mixture heated from below but in a rectangular geometry [21]. The fluid streaming upwards or downwards is indicated by white and black coding, respectively.*

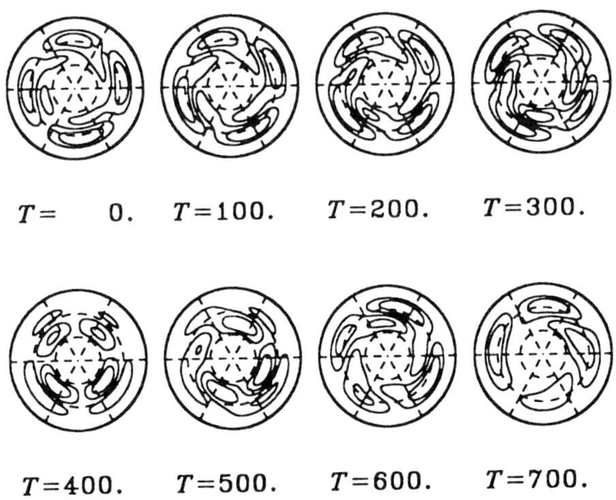

Figure 8.10: *Calculation of atmospheric patterns on the earth. Shown are results of our calculations [22] when looking at the top of the North Pole. The global features of weather maps are well caught by these calculations.*

spatial, spatio-temporal patterns, and also to chaos though space does not allow me to enter these questions here. At any rate I hope that thc reader has obtained the feeling that quite different systems may show the same behaviour because the equations (8.4.11) can be put into specific classes depending on the form of the nonlinear functions $\tilde{N}_u$.

We add two remarks in order to prepare the next section on pattern recognition.

1) Because the order parameters ξ_u are an order of magnitude larger than the amplitudes ξ_s of the enslaved modes, it is useful to split the decomposition (8.4.6) into a sum over the order parameters and a sum over the enslaved mode amplitudes according to

$$q = q_0 + \sum_u \xi_u(t) v_u(\mathbf{x}) + \sum_s \xi_s(t) v_s(\mathbf{x}) \, . \tag{8.4.13}$$

mode skeleton

The evolving structure is then primarily determined by the first sum which is called the mode skeleton.

2) In many important cases the right-hand side of (8.4.11) consists of polynomials up to third order where often the quadratic term vanishes because of symmetry reasons. In such a case the order parameter equations acquire the form

$$\dot{\xi}_u = \lambda_u \xi_u - \sum_{u'u''u'''} C_{u'u''u'''} \xi_{u'} \xi_{u''} \xi_{u'''} + F_u \, . \tag{8.4.14}$$

To mention two important examples: equations of type (8.4.14) govern the behaviour of multimode lasers where ξ_u can be identified with the field mode amplitudes of the electric field strength. The second example is provided by fluid dynamics where ξ_u is the order parameter belonging to the amplitude of a fluid roll in a fluid heated from below and (8.4.14) describes the competition between rolls where in each case only one mode wins the competition. In a number of cases of practical importance, eq. (8.4.14) can be written as potential equations:

$$\dot{\xi}_u = -\frac{\partial V}{\partial \xi_u} \, . \tag{8.4.15}$$

8.5 From pattern formation to pattern recognition

So far synergetics has been concerned with the formation of patterns in a great variety of systems. In this section I wish to show that pattern recognition can be interpreted as some kind of pattern formation. We shall base our procedure on three ideas: the first is that of associative memory which has been formalized by Steinbuch [23], Kohonen [24] and others. An example of an associative memory is provided by a telephone dictionary. Once a name is given, we may complete it by the corresponding telephone number. Quite generally, when a set of data is given, it will be complemented by an associative memory. For instance when a face or part of a face is given, it may be completed by adding the name of the corresponding person. In general, any object to be recognized is composed of so-called features. For our purposes it will be sufficient to exemplify our approach by assuming that the grey values in each pixel of a picture serve as these features (see below). Second we assume that in analogy to pattern formation, pattern recognition is done as a dynamic process in a synergetic system. The dynamics is described by order parameters which shall obey a potential dynamics in analogy to equation (8.4.15), a suggestion I had made about a decade ago [17]. Third, we adopt the basic idea of synergetics based on order parameters. In pattern formation the following happens: once an initial preparation or an initial fluctuation has generated a part of an ordered pattern of the system, that part generates the corresponding order parameter which in turn, via the slaving principle, forces the whole system into the specific totally ordered state. This behaviour of the system has its complete analogy with an associative memory: once part of a pattern is given, it generates its corresponding order parameter which in turn restores the total pattern. Let us consider the recognition of faces as an explicit example for our procedure. The faces are encoded as the grey values at each space point $\mathbf{x}$. The thus resulting description is represented by the function $v_u(\mathbf{x})$ where u distinguishes the different prototype patterns to be stored in the computer (fig. 8.11). When an incomplete pattern is offered it will be described by a function $q(\mathbf{x})$ which we may decompose according to (8.4.13) where we put $q_0 = 0$ and ignore the second sum. For a more detailed discussion how to take into account the second sum adequately in the pattern recognition process we refer the reader to the original literature.

In the following we shall be concerned with the dynamics inherent in the first sum because this will represent the most interesting part of our approach. The idea is now as follows: once an incomplete pattern q is given, it will be subjected to a dynamics which, eventually, pulls this initial pattern into a final pattern v_u depending on which pattern v_u is resembling the original pattern q the most:

$$q(\mathbf{x}) \to q(\mathbf{x}, t) \to v_u(\mathbf{x}) \,. \tag{8.5.1}$$

As I have shown all these requirements are met by the following set of equations [25]

Figure 8.11: *Example of patterns, in particular of faces, which were stored in the computer jointly with their labels indicating the names.*

$$\begin{aligned} \dot{q} &= \sum_u \lambda_{\mu} v_u(\mathbf{x}) < v_u^+ q > \\ &- B \sum_{u \neq u'} < v_u^+ q >^2 < v_{u'}^+ q > v_u \\ &- C(\sum_{u'} < v_{u'}^+ q >^2) \sum_u < v_u^+ q > v_u \, . \end{aligned} \tag{8.5.2}$$

The λ_u's are called attention parameters because they govern the selection of patterns in the recognition process, v_u^+ are functions adjoint to v_u obeying the equations

$$< v_u^+ v_{u'} > = \delta_{uu'} \, . \tag{8.5.3}$$

The bracket denotes integration over space. The second sum on the right-hand side of (8.5.2) takes care of the discrimination between patterns whereas the last term serves for saturation, i.e. the vector q remains bounded. When we multiply the expression for the mode skeleton in (8.4.13) by v_u^+ and integrate, we immediately obtain because of (8.5.3) the order parameters ξ_u

$$< v_u^+ q > = \xi_u \, . \tag{8.5.4}$$

This allows us to replace the dynamics inherent in q by the order parameter dynamics, and in this way we can transform (8.5.2) into

$$\dot{\xi}_u = \lambda_u \xi_u - B' \xi_u (\sum_{u'} \xi_{u'}^2) + C' \xi_u^3 \, . \tag{8.5.5}$$

(a)

(b)

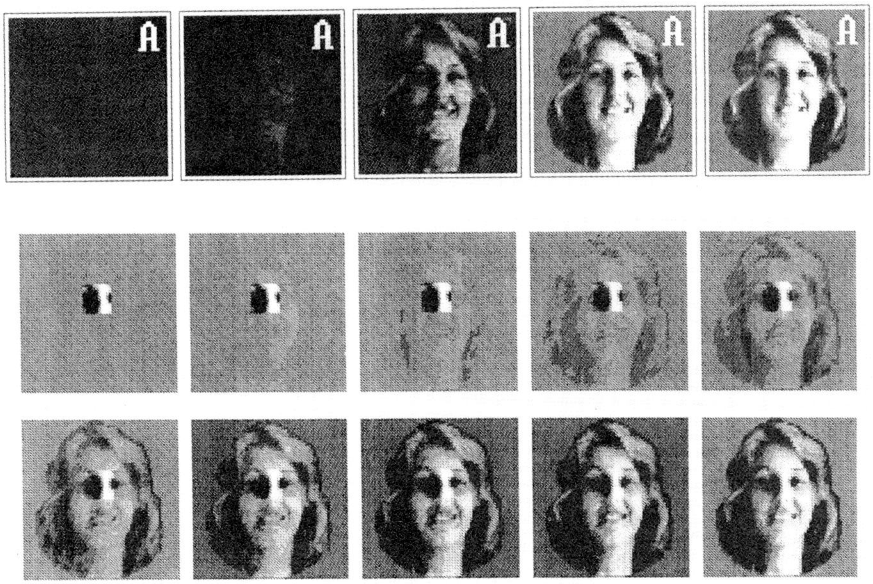

Figure 8.12: *(a) Restoration of a pattern from which initially only the letter A was offered to the computer. (b) In this case only a part of a face was offered to the computer which then was able to restore the whole face in the sense of an associative memory by a dynamical process.*

Quite evidently (8.5.5) is a special case of (8.4.14) which substantiates our claim that the dynamics of pattern recognition may be identified with that of pattern formation.

Let us illustrate out approach by a few explicit examples which are actually performed on a serial computer [26]. Fig. 8.11 shows a number of faces jointly stored with their names as prototype vectors $v_{\mathbf{u}}(\mathbf{x})$ in the computer. Then incomplete patterns as those giving only a letter encoding for the name or only part of a face were presented to the computer. By use of the dynamics (8.5.2) or equivalently (8.5.5) the computer could restore the total pictures (fig. 8.12). By adequate preprocessing of the pictures we could make the whole procedure invariant against simultaneous translation, rotation and scaling. Our procedure is capable of recognizing scenes. To this end we made the procedure invariant against translation and offered the computer a picture composed of two faces, one partly hiding the other one (fig. 8.13) [26]. Then the computer first recognized the lady in front. Then we put the attention parameter belonging to this lady equal to 0, and in the next run the computer recognized the man in this picture. In this way our procedure could decompose scenes composed of up to 5 faces.

Figure 8.13: *Example of a composite scene that was recognized by the computer (compare text).*

But I do not doubt that the procedure is good even if the faces are still more numerous so that group photographs can be recognized in this way. As one may show, our procedure or algorithm can be realized on various substrates or on various kinds of computers. In the actual work we did here, we used a usual serial computer. However, one may readily convince oneself that one can easily realize the whole procedure on a parallel computer in which each element has to perform multiplication and addition of incoming signals [25]. This kind of computer therefore is based on a construction principle different from the one typical neural nets are based on nowadays, namely on threshold elements that were originally introduced by McCulloch and Pitts [27] nearly fifty years ago and which have been utilized in spin glass models. Our model has certain advantages over the spin glass model, in particular it does not contain any spurious states. The minima of the potential function in which the dynamics of ξ_u takes place are precisely the positions of the stored prototype patterns.

8.6 Concluding remarks

This Symposium in honour of Prof. Casimir is concerned with the interrelation between science and technology. I hope that my paper may serve as a little illustration how their interaction may be very fruitful for each other. In the field of synergetics of which I gave a brief outline above, I started with a typical technical device, namely the laser. The theoretical study of this device led to what I believe are some rather deep insights into cooperative effects in systems far from equilibrium. This in turn allowed us to make some progress in science, namely to understand that pattern formation in a huge variety of systems ranging from physics over chemistry and biology to economics, is governed by few fundamental principles. And now we may return to technology, namely to the construction of new types of computers based on principles unearthed in pure science.

References

[1] C.J. Gorter and H. Casimir, Physica **1** (1934) 306; Z. Techn. Physik **15** (1934) 539.

[2] A.E. Siegman, *Lasers*, University Science Books, Mill Valley, California, 1986.

[3] H. Haken, *Light*, Vol. 2, *Laser Light Dynamics*, North Holland, Amsterdam, 1985.

[4] M. Sargent III, M.O. Scully and W.E. Lamb, Jr., *Laser Physics*, Addison-Wesley, Reading, Massachusetts, 1974.

[5] H. Haken, *Laser Theory*, Encyclopedia of Physics, Vol. XXV/2c, Springer-Verlag, Berlin; 2nd corr. ed. 1984.

[6] H. Haken, Z. Phys. **181** (1964) 96.

[7] H. Haken, Rev. Mod. Phys. **47** (1975) 67.

[8] J.A. Armstrong and A.W. Smith, Phys. Rev. Lett. **14** (1975) 68.

[9] H. Risken, Z. Phys. **186** (1965) 85.

[10] E.R. Pike, in: *Quantum Optics*, eds. M.S. Kay and A. Maitland, Academic Press, New York, 1970.

[11] F.T. Arecchi, G.S. Rodari and A. Sona, Phys. Lett. **25A** (1967) 59.

[12] C.E. Shannon, Bell System Techn. J. **27** (1948) 370, 623; Bell System Techn. J. **30** (1951) 50.

[13] H. Haken, Phys. **B61** (1985) 329; Z. Phys. **B62** (1986) 255; Z. Phys. **B63** (1986) 505.

[14] M. Schindel, Diplomarbeit Universität Stuttgart, 1987.

[15] H. Haken, *Information and Self-Organization*, Springer-Verlag, Berlin, Heidelberg, 1988.

[16] H. Haken and R. Graham, *Synergetik — Die Lehre vom Zusammenwirken*, Umschau **6** (1971) 191.

[17] H. Haken, *Synergetics, An Introduction*, Springer-Verlag, Berlin, Heidelberg, 3rd ed., 1983.

[18] H. Haken *Advanced Synergetics*, Springer-Verlag, Berlin, Heidelberg, 1987.

[19] R. Friedrich and H. Haken, Phys. Rev. **A34** (1986) 2100.

[20] M. Bestehorn, R. Friedrich and H. Haken, Z. Phys., to be published.

[21] M. Bestehorn, R. Friedrich and H. Haken, to be published.

[22] W. Weimer and H. Haken, J. Atmos. Sci. **46** (1989) 1207.

[23] K. Steinbuch, Kybernetik **1** (1960) 36.

[24] T. Kohonen, *Selforganization and Associative Memory*, Springer-Verlag, Berlin, 2nd ed., 1987.

[25] H. Haken, in *Computational Systems, natural and artificial*, Springer-Verlag, Berlin, Heidelberg, 1987.

[26] A. Fuchs and H. Haken, Biol. Cyb. **60** (1988) 17, 107, 476.

[27] W.S. McCulloch and W. Pitts, Bull. Math. Biophys. **5** (1943) 115.

Part V:
History of Industrial Physics

Between Science and Technology
A. Sarlemijn and P. Kroes (Editors)
Elsevier Science Publishers B.V. (North-Holland), 1990

Chapter 9

Innovation and basic research in the industrial laboratory: the repeater, transistor and Bell Telephone System

L. Hoddeson

University of Illinois

Urbana, Illinois

9.1 Introduction

How to model the relationship between the obviously intertwined phenomena of science and technology has been a question of great interest to Professor Hendrik Casimir[1]. In the concluding chapter of his autobiography, he presents four models: three "caricatures", which exaggerate one particular feature of the relationship, and his preferred more general model, "the science-technology spiral" [1]. His first two models — that science and technology are independent, and the standard myth that scientists, "feasting at the high table of the gods",

[1]For full references see the two articles on which this paper is based, L. Hoddeson, "The Emergence of Basic Research in the Bell Telephone System, 1875–1915", Technology and Culture **22** (1981) 512–544 (©1981 by The University of Chicago) and "The Discovery of the Point-Contact Transistor", Historical Studies in the Physical Sciences **12/1** (1981) 41–76 (©1981 by the Regents of the University of California). The editors thank the University of Chicago and the Regents of the University of California for permission to publish material from these articles.

are unconcerned about technology, but occasionally will "drop a few crumbs that are then gingerly picked up by lesser people, like engineers and captains of industry and generals and statesmen"— apply only in a limited number of cases. They are both opposed by the case studies to be analyzed here. Casimir's third caricature, in which "the course of technology and science is dictated by capitalist enterprise", which merely allows some free exploration as part of a general "policy of 'repressive tolerance'," seems also limited to a small class of examples, although it does fit superficially both the transistor and repeater. As Casimir notes, this model does not sufficiently express the intrinsic independence of the scientists who are in search of understanding. And, of course, it does not apply to non-capitalist economies; it cannot, for example, explain the early roots of science in China. The fourth model sees the paths of development of technology and of science as occurring in complex streams of advance that come into contact at a number of points and form a spiral, in which "technology uses scientific results but with a time lag", while "science uses technology without a time lag". This model is certainly the most general of the four, fitting every example that I have conjured up, but it is not yet sufficiently detailed to offer explanations, and it totally ignores the larger context in which science and technology are supported. For example, it ignores the strong causal influence that nonscientific concerns often have on both technology and science.

In this paper, I present two case studies of technological innovation, in the context of a science-based industry, that point to another model, a three-phase feedback loop in which initially, in the first phase, nonscientific objectives raise technological needs. In the second phase, these technological needs give rise to problems for which it is necessary to establish one or more basic scientific research efforts. And in the third phase, the technological success of these research efforts reinforces the company's decision to support such research [2]. In the case of the telephone repeater, the American Telephone and Telegraph Company (AT&T), the parent firm of the Bell Telephone System, had the goal early in this century of achieving a "universal telephone system". This non-scientific objective gave rise to the technological problem of developing an electronic amplifier, a problem so challenging that in 1911 AT&T established its first in-house research branch. The successful development of the repeater, a triumph that enabled transcontinental and transatlantic telephony, strongly reinforced the company's policy to support basic research. By 1925 the engineering department in which this research branch was housed had evolved into the Bell Telephone Laboratories. History repeated itself just after World War II when the telephone system's pressing need to replace the unreliable vacuum tube amplifier by a solid-state device led to the formation of groups devoted to basic solid-state research. When one of these groups developed the first transistor in late 1947, this technological success strengthened the company's support of basic solid-state research.

These two case studies also portray a second important theme that I wish to stress: the evolution within an industry of a kind of research environment that is particularly nurturing to the development of both science and technology. This environment resembles those found at the better research universities: researchers having high levels of academic training are given considerable research freedom and are encouraged to communicate their findings in journals and at professional meetings. A similar evolution occurred at the Philips Research Laboratory in Eindhoven, where Casimir worked from 1942 until his retirement in 1972, assuming the post of co-director of Philips in 1946, and acting as a member of the board of management from 1956 [3]. There were in fact fruitful interactions between the research leaders at the two laboratories; Casimir recalls visits to Philips by Mervin Kelly, who was Bell Laboratories' Executive Vice President and later President. Kelly and co-founder Anton Philips, agreed on the need for supporting fundamental research in the context of industry, even if the research showed no signs of ever being applied. Yet at both laboratories, practical interests remained a powerful guiding principle. The paradoxical fact was that at both laboratories the management judged the mission-oriented research to be most likely successful if this environment allowed the scientists to follow freely their own interests. Indeed, in both institutions there was always a counterpoint of attitudes that resulted in the particular research policy. At Philips, co-founder Gerard Philips, who had been deeply impressed by the filament and lamp work at the General Electric Research Laboratory in the United States, always viewed research as a means of reducing cost in the production of incandescent lamps. On the other hand, Gilles Holst, Philips' director of research, was guided in his design of the research organization by a strong interest in understanding as well as controlling materials. Out of a joint concern with the practical and fundamental, Holst shaped a laboratory with a research philosophy summarized in his "ten commandments" for industrial research, listed in Casimir's autobiography. Bell Laboratories came also to support this approach. I will trace in this paper the origins at Bell of the six following Holst commandments, which also define Bell Laboratory's "enlightened philosophy":

1. "Engage competent scientists, if possible young ones, yet with academic research experience."

2. "Give them a good deal of freedom and give a good deal of leeway to their particular preferences."
3. "Let them publish and take part in international scientific activities."
4. "Do not split up a laboratory according to different disciplines, but create multi-disciplinary teams."
5. "Give the research laboratory independence in choice of subjects, but see to it that leaders and staff are thoroughly aware of their responsibility for the future of the company."
6. "In choosing research projects, be guided not only by market possibilities but also by the state of development of academic science."

9.2 The telephone repeater

9.2.1 Early technical activities in the Bell System, 1875–1907

A survey of technical activities of the telephone company between its start in 1875 and AT&T's major reorganization in 1907, reveals several early roots of research, but no strong research commitment by the parent organization [4]. The Bell Company started with a funding and patenting agreement signed in February 1875, according to which businessman Thomas Sanders and attorney Gardiner G. Hubbard agreed to support Alexander Graham Bell's experiments to build a device for transmitting speech along wire in return for a share in any patents that might emerge. None of the three was a scientist. Bell himself, an amateur inventor, was a professor of vocal physiology concerned with teaching his father's system of visible speech to deaf-mutes; although he became quite knowledgeable in certain scientific areas, his primary concern was to apply science, not create it. Neither was Bell's technical assistant, Thomas Watson, a scientist. Their telephone experiments drew heavily on the science that both men had picked up, and received continuing input from available scientists. But their striking success, in March 1876, when Bell transmitted the famous first telephone message, "Mr. Watson, come here, I want you", was defined in practical not scientific terms.

When the telephone industry started in May 1877, the subscribers were encouraged to make their own electrical connections and string their own wires, but for an additional fee, the proprietors would carry out these jobs for them. As the telephone system expanded, the company assumed an increasing number of such technical engineering roles, and it would be out of these that the Bell System's basic research eventually developed. The documents show no clear-cut evidence of basic research in the late 1870s or early 1880s, but there are suggestions. For example, Emile Berliner had "charge of general experiments upon the transmitter and Magneto Telephone" and was expected "to make original investigation, in regard to speaking telephones generally", and "special experiments upon forms or modifications of Transmitters belonging to the Company" [5].

The character of the research began to change about 1886 with the hiring of the Harvard-trained research physicist Hammond Vinton Hayes, one of the first physics Ph.D.'s in America (and the second from Harvard). While Hayes' position allowed him little time for scientific study, he was both interested in and capable of carrying out basic research. A number of the engineering problems of the 1880s, while ostensibly preventing Hayes' department from pursuing basic studies, proved to be the source of Bell's soon-to-be-instituted policy of supporting basic research. Such problems included: transmitting current with maximum efficiency; developing switching (it was then still necessary for subscribers to "ring off" before an operator removed a connection); diminishing attenuation, interference and cross-talk; replacing the single-wire grounded circuits by two wire all-metallic circuits; and placing the wires in large cities in underground metallic circuit cables. Insoluble by cut-and-try methods, these problems required understanding electromagnetic principles still on the frontiers of physics, in particular, the theory of transmission lines operating at high (i.e. vocal) frequencies.

Transmission line theory was based on the principles of electromagnetic wave propagation, discussed in Maxwell's 1873 treatise on electricity and magnetism, but not entirely confirmed until Hertz's experiments in 1888. In the 1880s Lord Rayleigh and Oliver Heaviside began to apply Maxwell's theory to high-frequency transmission problems. But to apply Heaviside's and Rayleigh's theories required mathematical facility beyond that of anyone on Hayes' staff in the 1880s. Hayes began to express his belief that it might be profitable to hire a number of mathematically trained scientists who would address not only the immediate engineering problems but the "many problems daily arising in the broad subject of telephony which require solution but are not studied as they will not lead to any direct advantage to ourselves" [6].

Some of the electromagnetic problems of the 1880s were solved by men such as John Carty who were without formal scientific training but had ample experience in practical telephony. Carty, who had joined the company in 1879 at age eighteen as a switchboard operator, having given up his plans to go to Harvard due to a temporary vision impairment, proved to be an exceptionally gifted telephone engineer. One of his most important discoveries, published in 1889, was that, under many conditions, the disturbances responsible for cross-talk are electrostatic rather than electromagnetic; this result enabled him to develop a method for minimizing cross-talk by suitably intertwining lines at particular points called "neutral points".

By the 1890s, practical problems so overcommitted the Company that basic scientific studies were de-emphasized. In 1892 Hayes abandoned the research work of his department so that it could devote "all our attention to practical development of instruments and apparatus". He added, "I think the theoretical work can be accomplished quite as well, and more economically, by collaboration with the students of the [Massachusetts] Institute of Technology and possibly of Harvard College" [7].

But while focusing on "practical developments", Hayes also began to add a few university trained scientists to his staff. The first, hired in 1890 from Johns Hopkins University, was John Stone Stone, who applied his advanced mathematical training first to the theory of voice transmission and subsequently to amplifiers and Hertzian waves. In 1897, Hayes hired George Campbell, with a bachelor's degree in engineering from MIT in 1891, a master's degree from Harvard in 1893, a year at Göttingen studying with the mathematician Felix Klein, a year at Vienna studying with Boltzmann and a year in Paris studying under Poincaré. Campbell immediately turned to the principal problems at the turn of the century, reducing attenuation and phase distortion. These effects set a practical limit on "unloaded" lines of some 1,200 miles, a distance achieved in 1893 with the building of the line connecting Boston and Chicago. Extending the mathematical studies of Heaviside and Vashy, reported in 1887, that showed attenuation could be substantially reduced by uniformly distributing inductance along a line, Campbell, by 1899, developed the fundamental theory of the loading coil. This major contribution to telephone transmission involved connecting induction coils into telephone lines at definite intervals determined by the wavelength of the signal. In 1901, Campbell would take his doctorate in physics from Harvard with a dissertation based on his loading-coil studies [8].

It is worth noting, in passing, how Campbell's work on the loading coil provides yet another working example of the historical framework described in the introduction: the nonscientific goal of extending long-distance service gave rise to the technological problems of attenuation and phase distortion, which then stimulated Campbell's scientific investigation. His technological success would eventually increase the company's appreciation for the usefulness and importance of in-house basic research.

The third university-trained scientist Hayes hired before the turn of the century was the Canadian, Edwin Colpitts, who some years later would direct Bell's first research division.

Colpitts, hired in 1899, had a physics master's degree from Harvard taken in 1897 and two years of additional study in physics and mathematics, also at Harvard. Like Stone and Campbell before him, Colpitts was initially assigned to engineering problems, e.g., developing methods and instruments for measuring alternating currents, how to load phantom circuits and reducing inductive interference caused by electrical trains and trolley cars.

In 1904, Hayes hired as a transmission engineer another physicist, Frank Baldwin Jewett, then an instructor in physics and electrical engineering at MIT. By 1912, Jewett was assistant chief engineer at Western Electric; by 1921 he was vice-president and director; and in 1925 he became the first president of Bell Telephone Laboratories. Jewett's rise in the hierarchy exemplifies what would become an unwritten Bell policy of choosing research directors from the company's own technical staff. Association with leading physicists ran deep in Jewett's career. At the University of Chicago, where he took his doctorate in physics, Jewett had worked for A.A. Michelson and formed a close friendship with Robert Millikan, then a young physics instructor, who exposed Jewett to some of the recent developments in the new field of electron physics. Jewett would later draw upon these associations in helping to establish Bell's first research branch.

In the period 1900–1907, Hayes' reports continued to list work on problems dependent on electromagnetic theory, including analytical study of transmission on loaded long-distance lines, electrical interference, the phantom circuit and the multiple switchboard, while at the same time they played down science, and even original invention, by in-house staff. For example, in 1906 Hayes wrote to Frederick Fish, then president of AT&T: "Every effort in the department is being exerted toward perfecting the engineering methods; no one is employed who, as an inventor, is capable of originating new apparatus of novel design. In consequence of this it will be necessary in many cases to depend upon the acquisition of inventions of outside men, leaving the adaptation of them to our own engineers and to the Western Electric Company" [9].

How may we interpret Hayes' dual position regarding science and invention during the 1890s? One reasonable explanation is that in his official correspondence Hayes was responding to attitudes from above that considered science a somewhat risky investment. Hayes' pivotal role in the beginning of Bell's basic research program did not derive from his influence on policy but came, rather, from his having hired by 1904 the small core of academically trained scientists who in time would create the Bell System's first research division.

9.2.2 The transcontinental line and the first research branch

The company's stance toward in-house research changed drastically between 1903 and 1911, through a series of innovative steps initiated by business decisions. The company took the first of these steps in 1903 when it announced a new policy of developing a corporate image of scientific and technological superiority. At about the same time, Bell reversed its policy of withholding its research results from publication, and we find Campbell, for example, reporting in detail on his research in journals such as *Philosophical Magazine* (1903) and *Electrical World* (1904). But the most dramatic and influential changes began six months before the Wall Street panic in mid-October 1907.

The Bell System's financial position had become increasingly fragile in the decade and a half after Alexander Bell's original patents expired, in 1893 and 1894. Many independent telephone companies sprang up, and by 1900 there were over 6,000 companies. By 1907 almost half of the telephones in the United States were non-Bell. Subscribers were becoming increasingly dissatisfied with the service. The Bell System, having developed out of many different companies, was inefficiently and uneconomically organized. In April 1907, AT&T, finding itself in severe financial straits, underwent a management reorganization which brought a New York banking syndicate under J. P. Morgan into control of the company. Fish was replaced as president of AT&T by the assertive Theodore N. Vail, who had held that post twenty years earlier.

Vail immediately embarked upon a program designed to consolidate the Bell System and absorb the nationwide telephone system into Bell. Two of his early moves strongly influenced the development of Bell's program of research. The first was to join all ongoing research and development in the Bell System — consisting of AT&T's department of research and development in Boston and Western Electric's engineering departments in New York and Chicago. Some of the technical staff from Boston continued to function as the AT&T engineering department at AT&T headquarters in New York City; the rest were formed into a single centralized engineering department housed at Western Electric in New York City, the department that would evolve into the Bell Telephone Laboratories. Hayes was dismissed, and Carty, who had been Vail's associate in the 1880s, was appointed chief engineer.

Carty proved to be an eloquent spokesman for Bell's new policy of strongly supporting scientific research. His many writings were to be studded by the term "fundamental research", evidently one he enjoyed. And it is not therefore surprising that, starting in 1907, the engineering department's reports frequently would refer to "fundamental research", whereas earlier under Hayes' tenure this term had rarely been used. This rhetorical shift suggests the process that Vail set in motion in 1907 by placing Carty, a man who in time became a forceful science advocate, at the head of the new centralized engineering department. The move was to alter the way science was publicized in the company and create a supportive environment for the institutionalization of research.

Vail's decision in late 1908 – early 1909 to build a transcontinental telephone line also strongly influenced the development of research. He had set the broad goal, already referred to in the 1870s by Alexander Bell, of developing the Bell System into a communications network that would be capable of reaching "anyone – at any possible place", and publicized it widely as Bell's motto, "One policy, one system and universal service". Establishing coast-to-coast telephone service was an essential step. According to Vail's biographer, some Pacific businessmen had suggested to Vail that a line between New York and San Francisco be opened at the San Francisco Panama-Pacific Exposition, which was initially scheduled for 1914 (and later postponed to 1915). Although Bell economists were sceptical that such a line would be profitable, Vail, believing in the line's importance, authorized it anyway. Carty had concluded that the line would be feasible technologically.

But building the transcontinental line proved to be far more difficult than Carty expected. The technology for economically and accurately sending signals over distances of the order of 3,000 miles did not then exist, and the need to develop this technology would lead — as we shall now see — to the authorization of the company's first official research branch. The length of telephone lines had grown dramatically over the previous three decades. Unloaded lines spanned only 2 miles (between Boston and Cambridge) in 1876 but reached 900 miles (between New York and Chicago) by 1892. Energy loss was the chief limiting factor and set the practical extent of such lines at the 1,200 miles (between Boston and Chicago) achieved in 1893. Campbell invented the device by which this limit could be exceeded in 1900, and by 1911 the practical limit of loaded lines, the 2,100-mile distance from New York to Denver, was achieved by inserting loading coils every 8 miles, doubling the line's transmission efficiency. However, since loading coils do not replenish dissipated energy, any further increase in distance required a device, a "repeater", that would amplify attenuated speech waves.

In 1903–4, Herbert Shreeve, a member of Hayes' staff, developed a mechanical repeater that proved to be effective on relatively short lines (e.g. between Amesbury, Massachusetts and Boston) but was not adequate for long lines since it favored certain pitches over others and was highly distorting when used two or more in series; it was also disproportionately insensitive when the incoming signal was weak, and it failed entirely when connected into loaded lines. In 1910, Campbell suggested that the company look into the possibility of developing a repeater with a vibrating part consisting of mercury gas molecules or cathode rays (electrons). There is no evidence that this suggestion was approved. However, later that year, Jewett, on whom Carty had placed technical responsibility for the transcontinental line, wrote to the plant engineer at AT&T that the company needed "to employ skilled physicists who are familiar with the recent advances in molecular physics and who are capable of appreciating such further advances as are continually being made". He had demonstrated in a cost study that New York to San Francisco transmission without a suitable repeater would be uneconomical, but felt that "if this repeater matter is tackled in the proper manner by suitably equipped men working with full coordination and under proper direction the desired results can be obtained at relatively small cost" [10].

Considering that the solution of the repeater problem might lie in the area of microscopic physics, Jewett visited his old friend Millikan and explained AT&T's need for a telephone repeater. Jewett asked Millikan to send to AT&T "one or two, or even three, of the best young men who are taking their doctorates with you and are intimately familiar with your field. Let us take them into our laboratory in New York and assign them the sole task of developing the telephone repeater" [11]. Millikan sent several young Ph.D.'s to AT&T. The first, Harold Arnold, reported in January 1911.

In-house research became an official commitment of the Bell System in April 1911, when a special research branch was organized within the Western Electric Engineering Department to address the "increasing number of problems intimately associated with the development of the telephone business, which require especially exhaustive and complete laboratory investigation". A subgroup under Colpitts studied the repeater itself. It was specified, further, that "to make adequate progress" the new branch "should include in its personnel the best talent available and in its equipment the best facilities possible for the highest grade research laboratory work". The department had twenty members in 1912; by 1915 it had between forty and forty-five, at least seven of whom were Ph.D. scientists. The earliest problems of the new branch, while approached scientifically, were all directed toward solving specific engineering problems. In time the department would broaden its activities to include scientific studies having no immediate, but only possible future, applications. "Basic" industrial research was now recognized as intrinsically dual in nature, being fundamental from the point of view of the researchers while at the same time supported by the company for its possible applications.

On joining Bell's research staff, Arnold immersed himself in detailed investigation of the repeater problem. Searching for a lighter vibrating part to replace the mechanical diaphragm in Shreeve's model, he experimented in 1911 and 1912 with mercury vapor molecules. However, Arnold dropped this work abruptly in November 1912 because of an important company decision: to try to adapt to the repeater problem a device whose "vibrating part" consisted of electrons, the audion (or triode) invented by Lee de Forest [12].

Arnold developed theories which dictated optimal constants of the circuit elements, and he redesigned the tube to make best use of the space charge of electrons inside, employing a high vacuum, an oxide-coated filament, a more precisely placed grid and a new grid circuit. Arnold's "high vacuum thermionic tube" solved the repeater problem, thus providing the pivotal technology necessary for the transcontinental line; it was first used on a commercial circuit (at Philadelphia, on a New York to Baltimore line) in October 1913.

The 3,400-mile transcontinental line, built with Arnold's repeaters initially located at Pittsburgh, Omaha and Salt Lake City, and later also at Philadelphia, Chicago, Denver and Winnemucca, Nevada, was first spoken over by Vail, in July 1914. The achievement of the line was properly celebrated at its commercial opening on 25 January 1915, with well-publicized cross-continental conversations. The appeal of the transcontinental line increased the public's as well as the Bell System's appreciation of the enormous power of industrial research. And the success of the new repeater then stimulated further research into vacuum tube physics, including thermionic emission, electric discharges in gases and the dynamics of electronic and ionic flow, by associates of Arnold.

External factors also had influence on the evolution of the Bell System's research. Several other historical studies have emphasized the role of outside work in radio telephony, or "wireless", in the same period that Bell was developing the repeater for the transcontinental line [13]. The first, by the Federal Communications Commission (FCC), whose investigators in 1935–37 explored the start of Bell's research program, concluded that "the threatened invasion of their [Bell's] wire telephone industry by the emerging science of 'wireless' telephony" was the principal cause of "the present broad scope of the Bell System's research". But the historical sequence of Bell's radio activities in the period 1890–1914, which we now briefly summarize, does not support this interpretation, for the research broadened well before radio might have appeared "threatening".

Carty recognized the possibility that non-Bell wireless research might lead outsiders to the telephone repeater before Bell could achieve it and mentioned this possibility in a 1909 memorandum urging "vigorous work upon the development of a more powerful repeater". He argued:

> "Whoever can supply and control the necessary telephone repeater will exert a dominating influence in the art of wireless telephony when it is developed. The lack of such a repeater for the art of wireless telephony and the number of able people at work upon that art create a situation which may result in some of those outsiders developing a telephone repeater before we have obtained one ourselves, unless we adopt vigorous measures from now on. A successful telephone repeater, therefore, would not only react most favorably upon our service where wires are used, but might put us in a position of control with respect to the art of wireless telephony should it turn out to be a factor of importance [14]."

The FCC investigators largely based their argument of the outstanding importance of the wireless threat on this passage. But it is to be noted that here Carty is employing the idea of establishing control in the wireless area as an additional argument to strengthen his request for intensified research on the repeater; he is not arguing in particular for the support of radio research. Carty recognized in 1909 the possibility — which indeed history fulfilled and now allows alternate readings — that the key to solving both the repeater and wireless problems would be the same.

Carty kept a careful eye on progress in the radio field; Lloyd Espenshied recalls "a card catalog of developments". In the crucial years 1907–10, when the Bell System was establishing its new science policy and its first research division, the company was therefore quite capable of assessing its vulnerability to "threatened invasion" by wireless. That it did not consider itself to be in any imminent danger is evidenced by the fact that it took until July 1914 for Vail and the board of directors to decide to support a comprehensive program of radio research.

The point is that there was little technological basis for a wireless threat before late 1912 since the apparatus for radio and telephony were very different. Receivers of that period required high power levels, and long-distance radio waves therefore had to be transmitted at correspondingly high powers. But the existing (low-power) telephone technology could not modulate high-power radio waves with a speech signal, nor did the technology for receiving and faithfully amplifying modulated waves exist. With the application of the vacuum tube as an amplifier in 1912 it became possible to modulate at low power levels and then amplify to the level needed to achieve the desired distance. Between 1912 and 1914, the company had prudently taken steps to purchase the patent rights to the audion from de Forest, thus preventing any possible patent threat.

Bell's highly successful research into radio, starting in 1914, was yet another reinforcing strand in Bell's now well-established research program. Detailed investigation of radio techniques by R.V.L. Hartley, R.A. Heising and others, resulted in a working model of a vacuum tube system that became the basis of the Bell System's exciting transcontinental and transoceanic wireless test conversations in late 1915. These dramatic demonstrations not only captured the public's interest, but in fact brought Bell, as Vail publicly reminded Carty in a telegram on one of the occasions — "one long step nearer our 'ideal' — a 'Universal System'".

On New Year's Day 1925, a new corporation, the "Bell Telephone Laboratories", took over the engineering department of the Western Electric Company, which by now had over 3,000 persons working on its staff. The Bell System thereby formally institutionalized its "continuous program of research and development necessary to the progress of the Bell Telephone System", which, as we have seen, had been developing almost from the time the telephone was invented.

9.3 The discovery of the point-contact transistor

In the July 1945 reorganization of The Physical Research Department of the Bell Telephone Laboratories, two new features were apparent: an orientation towards basic studies in the new solid-state subfield of modern physics and an emphasis on multidisciplinary team research. The discovery of the point-contact transistor thirty months later by John Bardeen and Walter Brattain was the first significant achievement of the new approach.

9.3.1 Scientific background and institutional setting

In 1945 solid-state physics had only recently assumed an intellectual and institutional identify as a research field [15]. A rich body of data had been growing for more than a century out of studies in disconnected areas, including electrical and thermal conduction, optical and mechanical properties of solids, crystallography, magnetism and metallurgy. In addition many centuries of work by artists and craftsmen had produced a vast fund of empirical knowledge about the behavior of materials. But before 1927, none of the conceptual schemes proposed to organize this data could claim wide success; they would fit one set of observations but not others, and different theories often contradicted each other. A unified framework was lacking. A rapid succession of fundamental papers on the quantum electron theory of metals, the most fundamental written between 1926 and 1932 by Wolfgang Pauli, Werner Heisenberg, Arnold Sommerfeld, Felix Bloch and Rudolf Peierls, resulted in a skeletal quantum theory of solids that solved most of the fundamental puzzles of the last quarter century concerning ideal metals. By about 1933 this work had provided the necessary framework, and the pace of development of fundamental theory slowed down. Many of the earlier contributors switched to other fields. Political factors, notably Fascist policies, amplified the transition.

Solid-state physics now changed course. New theorists entering the field emphasized approximate calculations of the properties of real, imperfect solids. In the United States this emphasis influenced graduate students at the two principal centers for solid-state theory in the early 1930s, Princeton and the Massachusetts Institute of Technology. At Princeton, Eugene Wigner trained Conyers Herring, Bardeen and Frederick Seitz, three leading members of the first generation of American physicists who would refer to themselves as solid-state theorists. At MIT, John Slater educated the first large group of Americans to work on the quantum theory of solids, including William Shockley. Another group with a practical emphasis grew up in Britain around Harry Jones and Nevill Mott at Bristol. Academic efforts interacted with industry to start a field that has become the most heavily populated in physics.

World War II helped to restore rapid progress to solid-state physics by demonstrating the effectiveness of collaborative work by theorists and experimentalists on complex problems pertaining to radar and the atomic bomb, by supporting development of new methods of producing pure and perfect samples of materials on a large scale and by encouraging new computational and experimental techniques. The solid-state field became institutionalized. In the United States, the American Physical Society (APS) held a "Symposium on the Solid State" in January 1945. Two years later the APS established a division devoted to the physics of solids [16].

At Bell Laboratories, Mervin Kelly, the recently appointed Executive Vice President, took the new developments in solid-state physics, and especially semiconductors, into account in his postwar reorganization of Bell's Physical Research Department. Semiconductors interested industry because of their abilities to rectify electrical current and to alter their conductivities on addition of impurities or the application of heat, light, or electrical voltages. This last property made semiconductors useful as electronic switches.

Kelly had recognized the promise of semiconductor research during his tenure as Director of the Vacuum Tube Department (1928–1934).Vacuum tubes and relays underlay the telephone system's tremendous expansion in the 1920s and 1930s: through its ability to rectify and amplify, the vacuum tube enabled the extension of telephony to great distances; through its switching ability, the relay made feasible a complex interconnecting network. However, both devices had inherent limitations. Relays were slow; electron tubes, though fast, were expensive, wasteful of power, unreliable and bulky. Kelly was interested in exploring semiconductor devices as possible alternatives.

As a practical matter, Bell scientists had been studying solid-state devices, the various semiconductor "istors" — resistors, varistors and thermistors — that form the ancestry of the transistor. Brattain and Joseph Becker began to study copper oxide varistors as modulators in carrier frequency equipment; Richard Grisdale's research on varistors related to the reduction of clicks in telephone transmission; and Gerald Pearson's and J.B. Johnson's study of thermistors responded to Buckley's request for a device that would regulate repeaters on transcontinental cables more sensitively than the thermocouples then in use. These early semiconductor studies at Bell cross-fertilized one another.

The study of copper oxide varistors in the late 1920s and through the 1930s stimulated several attempts to build semiconductor devices. Charles Demarest, an officer in AT&T's Radio Division, learned about L. Grondahl's and P. Geiger's copper oxide rectifier, invented in 1926 at the Union Switch and Signal Company, a subsidiary of Westinghouse. Suspecting that this rectifier might be useful to the company, Demerest asked Russell Ohl, a member of his department, to examine its physics. Ohl found that copper oxide rectifiers are not useful at high frequencies. Then Maurice Long of Bell's Research Division, and also its Director of Educational Relations, heard Grondahl discuss the new rectifier at the Bureau of Standards. Experimenting with copper oxide samples that Grondahl gave him, Long realized that some characteristics of this device "were so similar to those of vacuum tubes that we ought to find out what made them tick". By 1929 he had convinced the company to hire Brattain to work with Becker on the copper oxide rectifier.

9.3.2 Quantum theory of solids at Bell Labs

Meanwhile, Bell physicists and physical chemists, were coming to recognize that the new quantum theory was essential to understanding solids and undertook study of quantum physics despite the lack of a formal program in solid-state theory at Bell. Some staff members took formal courses in quantum physics at Columbia University, then only a subway ride from Bell Laboratories. More often, the acquisition of quantum theory came from independent study of texts, like Slater and Frank's *Introduction to Theoretical Physics* (1933). The depression actually helped this self-improvement by creating the threat of layoff and competition to avoid it. The depression also provided extra study time on "layoff days". But Bell researchers in the early 1930s could not through self-study achieve the level of the graduates of Princeton's or MIT's solid-state programs. The natural step would have been to hire recently trained Ph.D.'s in solid-state physics. A hiring freeze between 1930 and 1936 made this resolution temporarily impossible.

When the freeze thawed, Kelly, now advanced to Director of Research, recruited several outstanding new graduates. In 1936 Dean Wooldridge came from the California Institute of Technology (Caltech) and Shockley from MIT; in 1939, James Fisk from MIT and Charles Townes from Caltech. In an organizational move in 1938, presaging his postwar program, Kelly placed Wooldridge and Shockley together with Nix in a research group under unprecedentedly loose supervision. Nominally they were authorized to explore basic questions of "electronic conduction in solids". But in effect they had investigative freedom. Kelly's recruitment of first-class quantum theorists and his decision to leave them unfettered in the study of solid state were major institutional changes. They made possible the establishment of the research program at Bell Laboratories that produced the transistor. While the group had independence in its choice of research problems, the company's interests were clearly indicated in the groups's official authorization: "This case covers fundamental research work on the solid state and it is expected that eventually it will aid in the discovery of new materials or methods of processing old materials which will be useful in the telephone business" [17].

In these years, 1936–40, Nix, Shockley, Wooldridge and others on the staff were concerned to teach themselves the latest advances in solid-state theory. They set up a study group and met for several hours each week, partly on company time, to review systematically recent texts in the quantum theory of solids, including those by Mott and Jones, Richard Tolman and Linus Pauling.

A third major line of research at Bell that proved important for the invention of the transistor aimed to understand and improve the operation of point contacts. In the late 1930s George Southworth and other members of Bell's radio group at Holmdel, New Jersey, were seeking new methods for the detection of very short (40 cm) radio waves [18]. Vacuum tubes, which in the 1930s had replaced the earlier cat's-whisker rectifiers, were insensitive at high frequencies. Southworth decided to reexamine point-contact detectors and constructed one first from parts found while rummaging in the famous second-hand radio market on Cortlandt

Street in lower Manhattan. He found them to be much superior to vacuum tubes in the high frequency range. This discovery prompted a large and continuing research effort on point contacts. The attempt to find materials that work best in point contacts led in turn to a program of studies of silicon.

Galena, the material traditionally used in cat's-whisker detectors, was not useful commercially; because of the non-uniformity of its crystals, one had to hunt around on them for spots at which rectification would occur. Ohl studied more than 100 other materials and found silicon to have the most sensitive point-contact detection properties. Sensitive, but again erratic; using the material then available commercially, he found that the contact between the silicon and the whisker sometimes rectified in one direction, sometimes in the other, and sometimes not at all. With Grisdale's help, Ohl attempted to make the silicon more uniform by purifying it through melting. They found that they needed special furnaces, and Ohl in August 1939 engaged Jack Scaff and Henry Theuerer, two of Bell's metallurgists to purify the silicon.

In carrying out this purification by melting in high vacuum, Scaff and Theuerer came upon the strange effect that the direction of rectification varied in an uncontrolled fashion from one silicon ingot to the next. They were producing what came to be recognized as either "n-" or "p-type" silicon, depending on whether the majority electrical current carriers are, respectively, electrons or "holes". The behavior remained a mystery until wartime research confirmed Scaff's conjecture that it arose from impurities in the silicon composed of elements in the third column (p-type) and fifth column (n-type) of the periodic table [19].

In September 1939, while cooling hot silicon ingots very slowly in an attempt to avoid cracking, Scaff and Theuerer accidentally produced a sample in which two parts rectified in opposite directions. As Ohl soon discovered, using an oscilloscope and neon lamp whose current passed through a chopper, this ingot contained a "pn junction", an interface within a crystal between a p-type and n-type region. He demonstrated the strikingly large photovoltaic effect of this junction in Kelly's office by shining a flashlight on it. This demonstration early in 1940 showed the effect to be more than ten times greater than they had been getting with normal photoelectric cells. This demonstration was the climax of Bell's prewar studies pertinent to the transistor. Under Kelly's watchful eye, the three lines of semiconductor research were coming together through the internal communications within the Laboratory.

At this point the war claimed the entire research effort of the Laboratory and delayed the unification of Bell's solid-state program for several years. However, the wartime diversion of research did provide a more favorable environment for the prosecution of research. Of the approximately 1500 military projects at Bell Laboratories during World War II, radar, their principal effort, stimulated extensive study of semiconductor phenomena and of the technology for producing crystals and devices. Other military projects also prompted work with and on semiconductors. For example, Pearson and his coworkers developed thermistors in thin films for night-vision devices.

Westinghouse, General Electric, Sylvania, Du Pont, the University of Pennsylvania and Purdue University, among other laboratories in the United States, and General Electric, British Thompson-Houston, the Telecommunications Research Establishment and Oxford University in Britain also rigorously pursued semiconductor research connected with radar. Intercommunication created widespread awareness of the importance of semiconductors (including a possible semiconductor amplifier) and of the fierce competition in the field likely to occur after the war.

Well before the war ended, Kelly began to plan to meet the competition. In 1943, he pointed out that until recently research at Bell had not interacted much with other technologies and industries, notably radio. Around 1937, however, Bell's research and development, stimulated by the military research on radar, began to move into radio and then into very high microwave frequency ranges and broader transmission bands. Thus, Bell entered a new phase in which "technical cross-fertilization and economic comparison" were important factors to be taken into account. According to Kelly, the key to success was large-scale effort in basic solid-state research. He set up a program on "Physics of the Solid State", whose authorizations explained that the work aimed at "fundamental study of the physical properties of metals and other solids, such as electrical, thermal, mechanical, diffusional and structural, particularly crystallographical".

With the strengthening of national commitment to research by wartime successes, those responsible to Bell for the running of its laboratories could afford further to liberalize their "enlightened" attitude toward research and offer greater freedom to the Laboratories' most talented scientists. Staff members were warmly encouraged to participate in the larger scientific community. The number of seminars, journal clubs and study groups increased after the war, and efforts were also made to provide further education for junior staff. These attitudes and circumstances, together with easy access to a larger staff of scientific and technical specialists in many fields, made the research environment for some Bell scientists superior to that offered at the best universities.

Kelly's reorganization of the Physical Research Department towards the war's end made solid state the leading subject of exploration in the new quasi-university environment. Three of the nine departments into which Kelly divided all of Physical Research — Solid State Physics under Stanley Morgan and Shockley, Physical Electronics under Wooldridge and Electron Dynamics under Fisk — were authorized to carry out basic solid-state studies.The Bell Laboratories mission lurked in the background; as Kelly put it in the 1945 authorization of the new Solid State Department, "the research carried out under this case has as its purpose the obtaining of new knowledge that can be used in the development of completely new and improved components and apparatus elements of communications systems". He noted further the "great possibilities of producing new and useful properties, by finding physical and chemical methods of controlling the arrangement and behavior of the atoms and electrons which compose solids".

The organizational chart for July 1945 further divided the Solid State Department into four subgroups whose specialties reflected Bell's continuing interests in solid state: magnetism; contacts and carbon microphones; dielectrics as well as physical chemistry of solids; piezoelectricity, crystals for oscillators and propagation of sound in solids. Each subgroup was designed as a multidisciplinary team — a balanced mixture of individual specialists — so that the group as a whole would have a spectrum of expertise. Such an organization, as Kelly observed, had proved itself at the big wartime research laboratories like MIT and Los Alamos. To adapt it at Bell would be both prudent and progressive. A fifth subgroup of the Solid State Department, devoted to semiconductors and directed by Shockley, appeared on the chart of January 1946.

Following certain crucial additions to the staff, Bell Laboratories was now a leading international center for theoretical as well as experimental solid-state physics. In the Fall of 1945 Kelly authorized Wooldridge to invite Herring, a top solid-state theorist and former student of Wigner's to join the new Physical Electronics group. Herring came to Bell for a trial of a few months and by spring 1946 had accepted a permanent position.

Fisk and Shockley convinced Kelly also to offer a position to Bardeen, whom they had known from their Cambridge days in the 1930s. Bardeen was already one of the outstanding solid-state theorists in the country; he joined the semiconductor subgroup of the solid-state group late in 1945. Since the Murray Hill Laboratories were congested in 1945, Bardeen had to share an office with Pearson and Brattain, who were again studying rectifiers, varistors and thermistors. Bardeen was soon trying to explain the data that Pearson and Brattain were gathering on semiconductors.

9.3.3 Steps to the invention of the point-contact transistor

Shockley's semiconductor subgroup represented a range of pertinent specialties. Shockley himself and Bardeen were theoretical solid-state physicists; Brattain and Pearson, experimental physicists who had worked for more than a decade on semiconductors; Robert Gibney, an experienced physical chemist; and Hilbert Moore, an electronics expert. The team also had two technical assistants, Thomas Griffith and Philip Foy.

The first task was to review the wartime progress. Advances made at Bell during the war that would be crucial to the invention of the transistor include the identification by Ohl and others of the impurities that cause n- and p-type behavior and the development by the metallurgists Scaff and Theuerer of techniques for "doping" silicon and germanium. Seitz and coworkers at the University of Pennsylvania and Du Pont, studying silicon, and Karl Lark-Horovitz' group at Purdue, working on germanium, had significantly deepened understanding of the properties and methods of producing both high purity and doped silicon and germanium. The Purdue group had also developed the "high-back-voltage germanium" rectifier, which overcame the low-voltage limitations of the earlier crystal rectifiers and had improved rectification characteristics. Unlike the ordinary crystal rectifiers of that time, which at high voltages not only suffered physical damage but conducted in the reverse direction (their back resistance decreasing with increased negative voltage), the high-back-voltage rectifier of super-pure germanium had exceptionally low conductivity in the back direction for applied voltages up to about 100 volts.

Substantial progress had also been made by Mott, Walter Schottky and B. Davydov in 1938, and by Schottky, Hans Bethe, Robert Sachs, Karl Herzfeld and others during the war, in developing a theory to explain rectification at a junction between semiconductor and a metal, or between two different semiconductors. Both the early theory of Wilson, Lothar Nordheim, Jakov Frenkel and A.F. Joffe, and the later theory of Mott and Schottky attributed the rectification to the experimentally observed potential barrier at the interface, which was assumed to be caused by a region deficient in electrons, the "space charge rectification layer". In both the Wilson and Mott-Schottky models, the space charge layer functions as a rectifier in much the same way as does the region in a vacuum tube diode between the filament and plate.

One shortcoming of the theories of Mott and Schottky and of Wilson, was their focus on the majority carriers in the semiconductor (in our example, the electrons); both models ignored the minority carriers. Davydov took the latter into account in 1938, but his mathematical treatment masked this point, which received little attention at the time. Excluding minority carriers is indeed valid for copper oxide, where their flow is insignificant, but does not do for germanium and silicon. The Bell Laboratories semiconductor group would later learn, essentially by accident, that minority carrier flow is the key to the operation of the germanium point-contact transistor.

Most of the early amplifier experiments at Bell were based on the simplest analogy between the semiconductor and vacuum tube diode and consisted of trying to extend the analogy by inserting into a semiconductor diode a third element, which, it was hoped, would play the role of the grid in the vacuum tube triode. Brattain and Becker had tried to do so with copper oxide in the 1930s. The hitch, as they perceived it then, was lack of technique for placing a grid in the appropriate place in the barrier region of 10^{-4} cm. By computing the size of the grid needed to fit into this space, Brattain had concluded that there could be no feasible geometry.

The Mott-Schottky theory also stimulated the invention of amplifier designs. Shockley recalls being inspired to think along such lines at the time he was hired by a "lecture or pep talk" from Kelly on Bell's need to replace metal contacts (as in relays) by electronic switches [20]. Around 1939 Brattain and Becker interested Shockley in the copper oxide rectifier, and shortly afterwards Shockley conceived a way to employ the space charge rectification layer, as discussed by Mott and Schottky, as a valve to control conduction and achieve amplification. Shockley described the design, which would now be called a "Schottky-gate-field-effect" transistor, in his laboratory notebook on 29 December 1939; the basic idea was to create a region of high resistivity in the way of the current in the circuit of interest and electrically bias this region so that it would regulate current much as the biased grid does in a triode.

Wooldridge recalls watching Shockley experiment with such a device, provisionally constructed using two copper wires and a heavily oxidized piece of copper screen. "He was already then trying to put some batteries across there and see if he could influence the current flowing by the potential across the third electrode" [21]. The experiments did not reveal the predicted effect. But working with copper wires, metal-semiconductor junctions and batteries connected across pairs of these elements in an attempt to produce amplification, Shockley had before him in late 1939 the basic elements of a transistor.

Other materials besides copper oxide were employed in early solid-state amplifier experiments at Bell. In 1938 or 1939, Shockley had tried in collaboration with Alan Holden, to build an electro-mechanical amplifier using carbon contacts brought together through pressure, controlled by a signal applied to a piezoelectric quartz crystal. They expected, but did not get, a usable output through a change in the resistance of the carbon as a signal was applied. Ohl experimented with silicon. During the war he tried to insert a third element in a diode composed of two silicon surfaces, but the device could not develop a sufficiently low impedance because of its high capacitance. In another wartime experiment, Ohl built a device for amplifying radio signals that employed point-contact detectors having negative resistance created by thermistor heating. This device worked, but its marginal stability made it unsatisfactory for commercial use [22].

While none of the prewar attempts to build commercially usable semiconductor amplifiers succeeded, they identified relevant problems. In its effort to solve them, the Bell semiconductor group decided to focus on silicon and germanium, not only because they were much simpler than the copper oxide and selenium studied earlier, but also because wartime advances in purifying and controlling impurities in these materials made it possible to obtain consistently behaving samples. Germanium had been used extensively during the war in high frequency detectors; silicon, because of its larger bandgap, had found fewer uses but was recognized as "a thing of the future".

Characteristically, all members of the group took part in such a decision; indeed at no point in the studies leading to the transistor did any of them work alone for an extended time. There was almost daily informal discussions on all aspects of the work, as well as frequent interaction with outside specialists, including Scaff and Theuerer who supplied the silicon and germanium.

The road to the discovery of the point-contact transistor was opened in April 1945, when Shockley sketched the design of a "field-effect" amplifier in his laboratory notebook [23]. Shockley argued that if the contact potential field of a rectifier could produce a space charge layer at the interface between a metal and a semiconductor, with most of the potential drop on the semiconductor side spread out over a distance of the order of 10^{-4} cm, then an externally applied electric field should be able to create such a barrier region. Furthermore, if the semiconductors were sufficiently thin, changes in the applied field could significantly alter the number of charge carriers and modulate the current passing through the semiconductor, thus providing the basis for an amplifier. Once again, however, experiments to test the design, did not succeed. Some important, as yet unknown, phenomenon was being ignored.

Shockley continued to ponder the field-effect amplifier, and on 23 June 1945 calculated that in order for the effect not to be observed, the changes in conductivity had to be less than 1/1500 of that predicted by the Mott-Schottky theory. Sometime during the fall of 1945 or winter of 1945–46, he took his calculation to Bardeen, who verified its accuracy and highlighted the disparity between the existing semiconductor rectification theory and the experiments.

By the afternoon of 18 March 1946, Bardeen had developed a new theory that explained the failure of the initial field-effect experiments. This theory added to the existing model of semiconductor rectification the general postulate that electrons at the surface of a semiconductor can be trapped in "surface states". These trapped electrons do not participate in the conduction process in the semiconductor and so reduce the carriers available for the field effect. Furthermore, as Bardeen showed, if enough surface states are filled, the sheet of surface electrons forms a barrier, or space charge layer, at the free surface of the semiconductor similar to that at the interface of a metal and semiconductor. This layer electrically shields the semiconductor from the influence of the applied electric field on the control plate, thus cancelling most of the effect of the external field and greatly weakening the field effect. The failure of Shockley's initial field-effect design had led to a new physical theory, and this theory, by providing an explanation for the failure, renewed the group's confidence that they might achieve the device if they could learn enough about the nature of the surface states.

9.3.4 Return to field effect experiments

The direct line of crucial experiments began with one carried out by Brattain between 13 and 17 November 1947. While studying the change of the contact potential on illumination with light over a range of temperatures, he noticed that incidental condensation of water on the surface caused considerable hysteresis as the apparatus was taken from high to low temperatures and back again. In an attempt to avoid the hysteresis, which he regarded as a nuisance, Brattain decided to immerse the entire system in various electrolytes and dielectric liquids, including acetone, alcohol, toluene and distilled water. To his surprise, the photovoltaic effect increased; the mobile ions in the liquids, being very close to the surface, created an electric field strong enough to overcome the shielding of the surface states.

Brattain showed the results to Gibney, who on 17 November 1947 made the important suggestion, to "vary the DC bias on circuit while observing the light effect". (In Brattain's experiment this electrical bias between the semiconductor surface and the reference electrode had been used to counteract the contact potential.) On trial, the two noticed that varying the applied voltage caused a corresponding variation of the photovoltaic effect in cases where the

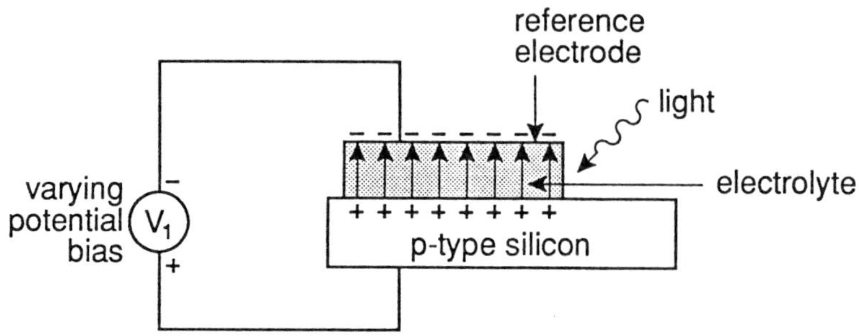

Figure 9.1: *Gibney's suggestion.*

liquids used in the experiment were electrolytes. They thought that the effect might be used in the design of an amplifier. (The role of the electrolyte was to provide ions at the surface of the semiconductor.)

The physical phenomenon associated with Gibney's suggestion can be understood from figure 9.1. Brattain was measuring the contact potential by applying a negative potential bias, V_1, to the reference electrode. The negative electrode attracted positive ions in the electrolyte and repelled negative ions, which, collecting near the surface of the sample of p-type silicon, caused positive charge in the silicon to come to the surface and fall into the positive surface states. The maneuver reduced the number of mobile carriers in the silicon. A very strong electric field existed at the surface owing to the double charge layer. The effect was measured by a photovoltaic experiment: light produced electron-hole pairs in the silicon; the positive surface charge repelled the holes and attracted the electrons; the high electric field at the surface fell, causing a measurable "photovoltaic impulse". Increasing the external field increased the impulse.

Gibney's suggestion brought the group back to the influence of applied fields on the electrical properties of semiconductors, back to an analogy to Shockley's field-effect experiments. Gibney and Brattain realized on 20 November that if they could project the change in the number of mobile charge carriers in the silicon that they were causing by altering V_1 onto a suitably arranged second circuit, rather than merely observing it as a photovoltaic effect, they might be able to build a field-effect amplifier. After almost two years of probing into the nature of the surface states and mobilities, the group appeared to be nearing its goal.

During this period Shockley continued to offer suggestions and explanations about the field effect. However, as a consequence of a visit to European solid-state laboratories in the

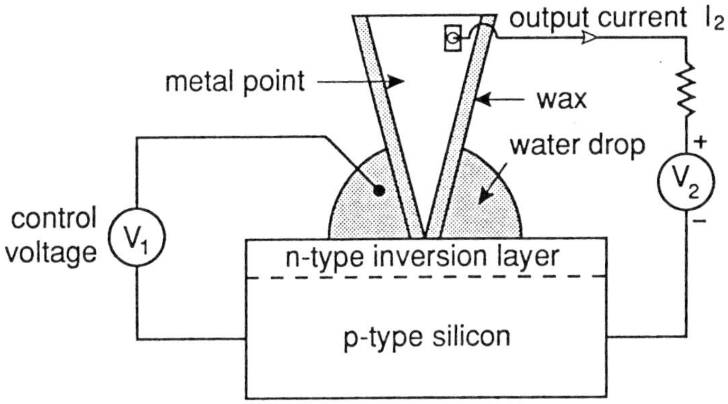

Figure 9.2: *Bardeen's proposed field-effect amplifier, 21 November 1947.*

summer of 1947, particularly Mott's in Bristol, he had become deeply involved in the theory of dislocations and the flow of electrons through alkali and silver halides. He did not participate directly in the field-effect experiments.

It was Bardeen who suggested how to add a second circuit that would convert the arrangement shown in figure 9.1 into a field-effect amplifier. Bardeen's proposal, shown in figure 9.2, employed a metal point contact and an "inversion layer", a thin region (discussed earlier by Schottky) in which the majority carriers are opposite to that in the bulk, which in certain circumstances forms near the surface of a semiconductor. The control voltage circuit here was the counterpart of the earlier circuit (fig. 9.1) that applied a bias V_1 to the reference electrode, and the electrolyte was a drop of distilled water. The current in the output circuit (I_2,V_2) passed through the metal point, which was pushed down on an n-type inversion layer on a p-type silicon sample; the point was insulated by a thin layer of wax and thus separated from the electrolyte. Here again, as in Shockley's earlier field-effect design, the control potential V_1, acting between the water and the silicon, changed the number of carriers in the silicon by attracting negative charge from the silicon top the surface states caused by the positive water ions at the surface of the silicon.

As Bardeen recalls, he suggested using a point-contact device at this time simply for "convenience"; he did not intend to go to high frequencies, or to create the ideal arrangement. Considerable art and understanding had been developed in working with point contacts since the return to this device almost a decade earlier by Southworth; a point-contact amplifier design could now be set up and tested in a single day. By using an inversion layer in place of the thin film of semiconductor of Shockley's earlier design, Bardeen hoped to take advantage of the large changes in conductivity that are possible in films, circumventing two problems inherent in film experiments: the practical difficulty of depositing a thin layer of semiconductor and the problem of the low mobility of carriers in such films. Mobility is considerably higher in an inversion layer than in a film. And Brattain knew that an inversion layer could be formed on the surface of silicon.

The experiment consisted of varying the applied or input voltage V_1 and measuring the resulting change (due to the change in the number of carriers in the silicon) in the output voltage V_2. On Friday 21 November, using this apparatus, they achieved amplification of both current and power, but not of voltage. Bardeen made a note on 22 November: "These tests show definitely that it is possible to introduce an electrode or grid to control the flow of current in a semiconductor".

The device was not yet usable: the water drop tended to evaporate during experimental tests and the apparatus responded only to low frequencies. During the next two weeks they tried many modifications. Gold and tungsten were used as electrodes and coated with Duco lacquer as well as with wax; for the electrolyte, besides water, tetramethyl ammonium hydroxide was tried and, at Moore's suggestion, various gels. On 26 November Brattain reported that gelatin in water "showed some promise but dried out"; they achieved better results with a mixture — they called it "gu" — of glycol borate and glycol bori-borate. Several geometries were considered, the most promising one having, in place of the water drop, a ring of gold on the "gu" with the metal point, as before, passing through the center of the electrolyte. Other configurations included a drop of the electrolyte on a junction of a p- and an n-type sample, two metal points placed close together on the n-type inversion layer on a p-type sample and a sandwich of germanium on SiO_2 on top of the n-type silicon inversion layer. Brattain noted on 8 December that, during a luncheon discussion, Bardeen had suggested replacing the silicon by the "high-back-voltage" germanium studied by the Purdue group. Since this high purity germanium made better rectifying contacts than ordinary silicon or germanium, they guessed that its use would improve the voltage amplification ratio. They soon found that an inversion layer would form on the special germanium, which they then used in place of silicon.

On 8 December 1947, with a gold ring electrode on a circular glob of the glycol borate mixture on the special germanium, and the metal point contact passing through the center of the ring, Bardeen and Brattain found a voltage amplification of two and a power amplification of 330. The frequency response remained poor owing to the glycol borate; the device failed to

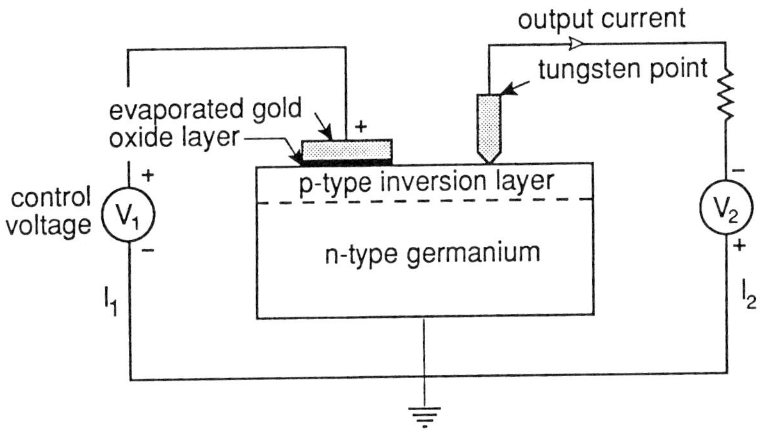

Figure 9.3: *Bardeen and Brattain's field-effect model of 11 December 1947, in which accidental contact occurred.*

amplify at frequencies much above 10 Hz.

Then, accidentally, working with the glycol borate and high-back-voltage germanium, Bardeen and Brattain observed an unexpected effect of great importance: they found that with a steady bias on the electrolyte, the glycol borate would etch the surface of the germanium, forming, as Gibney pointed out, an oxide film. They speculated that this accidentally grown oxide film between the electrolyte and semiconductor might, if insulating, act as a condenser and, since it was very thin, enable them to obtain a very large field with a moderate voltage and so increase the degree of modulation of the current flowing through the inversion layer. In effect the film would replace the glycol borate and might give a higher frequency response.

In the next experiment, performed on 11–12 December 1947, the currents were reversed by the replacement of a p-type sample by an n-type sample (fig. 9.3). An oxide replaced the drop of water. And to serve as the voltage plate in the control circuit, Gibney carefully evaporated a circular spot of gold upon the somewhat thicker oxide film, which had been grown on the germanium using a steady applied voltage. He left a hole in the center of the gold to allow the metal point to contact the germanium as in the previous configuration; but an electrical discharge between the point and the gold spoiled the hole, and Brattain resorted to placing the point at the edge of the gold, thus physically separating the two neighboring circuits, as the figure indicates.

9.3.5 The point-contact amplifier

Once again the elements for observing the field effect were present. However, the experiment did not work as planned, for the oxide layer did not insulate. To their great surprise on 11 December 1947, Brattain and Bardeen observed modulation of the output current and voltage in this configuration. But it occurred only when the gold was biased positively, i.e. oppositely to what was expected! According to the earlier arguments (cf. fig. 9.1) based on the supposition of a field effect, if the positive bias V_1 on the gold electrode increased, then holes would be driven away from the surface and out of the p-type inversion layer, decreasing I_2. However, the current I_2 in the output circuit, biased in the reverse direction, increased rather than decreased as the positive bias on the gold increased. This indicated that holes — the minority carriers in the bulk material — were flowing into rather than out of the inversion layer through some accidental contact with the positively biased gold. On realizing that the grid emitted holes, they renamed the control circuit the "emitter" and the output circuit through the tungsten point the "collector"; the corresponding currents were designated I_e and I_c respectively.

The attempt to use the oxide layer as a field-effect capacitor had therefore failed. But growing the oxide film and treating it with glycol borate had helped to produce the large inversion layer at the surface needed for the effect. The device gave some voltage amplification but no power amplification. It showed Brattain and Bardeen that one could build an amplifier on a principle entirely different from the field effect, that of point-contact.

The Bell Laboratories group was not the first to observe hole injection. In 1947 a group at Purdue including Ralph Bray, who had studied spreading resistance in the high-back-voltage germanium rectifier during the war, noted that the resistance increased with voltage when the rectifier was biased in the forward direction. This effect, arising from the injection of holes into the germanium, was the key to the point-contact transistor, but the Purdue researchers did not initially understand its physical basis and could not, therefore, use the phenomenon in the design of an amplifier. The Bell group became aware of Bray's experiments before they had completed their crucial experiments. As Bardeen recalls, they realized that the conductivity increased near the point by the injection of holes into germanium and became concerned that Purdue might beat them to what turned out to be their discovery of the first transistor.

The team had yet to produce a usable amplifier — a device in which the change in the collection current not only followed but exceeded the change in the emitter current. They

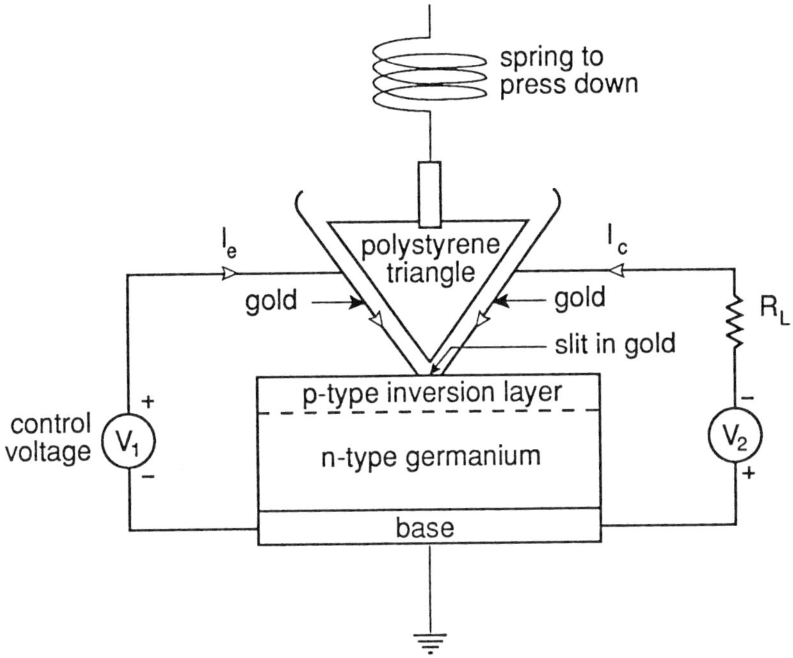

Figure 9.4: *The first transistor.*

had somehow to make the hole current flowing from emitter to collector to alter the barrier layer at the collector so as to make it easier for electrons to flow in the collector circuit. The climax of Brattain and Bardeen's experiments employed Bardeen's suggestion that they could perhaps obtain greater amplification if two contacts on the germanium were placed only a few millimeters apart. They believed that they could get a stronger effect if they used line rather than point contacts, but it later became evident that point contacts would do if enough current flowed in the collector. Brattain solved the technical difficulty of getting the line contacts close enough by wrapping a piece of gold tape around the tip of a triangle of polystyrene, cutting a slit on the gold at the apex with a razor, and filling the cut with wax. The separation was about $4 \cdot 10^{-3}$ cm.

The classic experiment, shown in figure 9.4 and differing from the arrangement in figure 9.3 only in the geometry and constitution of the contacts, worked the first time Bardeen and Brattain tried it, on 16 December 1947. In one of the first experiments at 1000 Hz, the power gain was 1.3 and the voltage gain fifteen. Thus was the point-contact transistor born.

After a week of checking their results, Bardeen and Brattain showed them to executives at Bell Laboratories. On 22 December 1947, they connected the input to an audio signal and the output to an oscilloscope. Brattain reported in his notebook on 24 December 1947: "This circuit was actually spoken over and by switching the device in and out a distinct gain in speech level could be heard and seen on the scope presentation with no noticeable change in quality. By measurements at fixed frequency in it was determined that this power gain was the order of a factor of 18 or greater".

The transistor was named by John Pierce in Brattain's office. Brattain recalls:

> "[We] wanted a name that would fit with "thermistor" and "varistor". Pierce first mentioned the important factor in vacuum tubes that was called "transconductance". Pierce knew that the point contact amplifier was the electrical dual of the vacuum tube. After some thought he then mentioned the dual of transconductance which is "transistance" and then he said "transistor" [24]".

The discovery was kept "laboratory confidential" for seven months, for Bell was concerned that the breakthrough might be occurring simultaneously elsewhere, perhaps at Purdue. Meanwhile, Bardeen worked with an attorney to draw up patents. The classic paper announcing the discovery was sent to the *Physical Review* with a request to hold back publication until Bell had received word that the discovery would not be classified.

On 30 June, after that decision was made, news of the discovery was released to the press.On 1 July *The New York Herald Tribune* reported, "The device is still in the laboratory stage but engineers believe it will cause a revolution in the electronics industry!"; *The New York Times*, mentioning the discovery in a small item in their "News of Radio" section, reported: "A device called a transistor, which has several applications in radio where a vacuum tube ordinarily is employed, was demonstrated for the first time yesterday at Bell Telephone Laboratories ... where it was invented".

From this point, Bell Laboratories engaged in a major effort to exploit and develop the transistor. The original transistor ultimately led to new families of solid-state amplification devices (which would soon eclipse the point-contact transistor's technological significance) and to the creation of whole new industries. It contributed significantly to the mushrooming of basic research in solid-state physics during the 1950s and 1960s, both at Bell Laboratories and in the larger scientific community.

9.4 Conclusion

As mentioned in the introduction, a simple way to schematise the relation between science and technology in the histories of the repeater and transistor is the three-phase feedback loop model, in which, in the first phase, non-scientific objectives opened particular technological problems whose urgency was fostered by external economic, social and political contexts. In the second phase, these technological problems, so profound or complex that the usual approaches to such problems failed, led the Bell System to seek deeper understanding of the underlying physical phenomena. To do so, they established a program of basic research, which, in the third phase, because of technological success, reinforced the company's commitment to research in the particular area. In the case of the repeater, the problem of extending telephone service to transcontinental and transatlantic distances led AT&T to establish its first in-house research department. Technological success reinforced the company's decision to support in-house research. Thirty-five years later, the pressing need to replace the vacuum tube amplifier by a solid-state device helped to motivate Bell Labs' decision to set up its basic solid-state research program. This program resulted in the first transistor, a success that strongly reinforced the company's decision to support basic solid state research.

This feedback loop model is by no means the only one that fits the data. Casimir's science and technology spiral works too and has greater generality, if less structure. In closing I would like to put forth yet another model, which admits generalization to contexts other than technology-based industry. We picture the relationships as a flow of influence among three spheres of activity: the nonscientific, the technological and the scientific. In the two cases treated here, we observed flow originating in the nonscientific sphere, streaming to the technological and finally to the scientific, with feedback from the scientific to the nonscientific via the technological. The more general historical problem of studying the relationships between science, technology and society becomes that of determining the influence flow pattern for each particular context. Out of an array of such flow patterns, a small corner of which this paper has filled in, historians may begin to identify the mechanisms underlying the more general relationships between science, technology and society.

References

[1] H. Casimir, *Haphazard Reality: Half a Century of Science*, Harper and Row, New York, 1983, pp. 294–313. Significant examples from the critical literature on the science-technology relationship are: R. Multhauf, "The Scientist and the 'Improver' of Technology", Technology and Culture **1** (1959) 38–47; D. Gruender, "On Distinguishing Science and Technology", Technology and Culture **12** (July 1971) 456–63; and almost the entire October 1976 issue of Technology and Culture, particularly T. Hughes, "The Science-Technology Interaction: The Case of High-Voltage Power Transmission Systems", pp. 646–59; O. Mayr, "The Science-Technology Relationship as a Historiographic Problem", pp. 663–72; E. Layton, "American Ideologies of Science and Engineering", pp. 688–70.

[2] Among the good studies of this subject are L.S. Reich, *The Making of American Industrial Research. Science and Business, GE and Bell, 1876-1926*, Cambridge, New York, 1985; Reich, *Radio Electronics and the Development of Industrial Research in the Bell System*, Ph.D. thesis, Johns Hopkins University, 1977; M.D. Fagan (ed.), *A History of Engineering and Science in the Bell System: The Early Years (1875-1925)*, Bell Laboratories, Warren, N.J., 1975; S. Goldberg, "Basic Research in the Industrial Laboratory: The Case of Bell Telephone Laboratories" (unpublished MS).

[3] Casimir, 1983 [1], pp. 224–54.

[4] Most of the references on the early history of the telephone and telephone system pertinent to this study are contained in R.V. Bruce, *Bell: Alexander Graham Bell and the Conquest of Solitude*, Little, Brown and Co., Boston, 1973; Fagen, 1975 [2]; P.C. Mabon, *Mission Communications: The Story of Bell Laboratories*, Bell Laboratories, Murray Hill, N.J., 1975 and *Events in Telephone History*, AT& T, New York, 1974.

[5] T.N. Vail to Lockwood, "Electrical Department", 5 Feb. 1881, Bell Laboratories Collection (herafter cited as BLC).

[6] Hayes to Hudson, 26 Oct. 1887 (BLC).

[7] Hayes to Hudson, 7 Mar. 1892 (BLC).

[8] N. Wasserman, *From Invention to Innovation: Long-Distance Telephone Transmission at the Turn of the Century*, Johns Hopkins University Press, Baltimore, Maryland, 1985.

[9] Reich, 1977 [2], p. 20.

[10] Jewett to Gherardi, 6 Dec. 1910 (BLC).

[11] O. Buckley, "Frank Baldwin Jewett, 1879-1949", National Academy Biographical Memoirs **27** (1952) 239–64, on pp. 247–48.

[12] G. Tyne, *Saga of the Vacuum Tube*, Howard W. Sams and Co., Indianapolis, 1977, pp. 57–61; J. Mills, "The Line and the Laboratory", Bell Telephone Quarterly **19** (Jan. 1940) 5–21.

[13] E.g., Goldberg and Reich [2]; N.R. Danielian, *AT&T: The Story of Industrial Conquest*, The Vanguard Press, New York, 1939 and the reports of the Federal Communications Commission, *Reports on Telephone Investigation (Pursuant to the Public Resolution No. 8, 74th Congress)* (House Document #340, 76th Cong. 1st sess. 1939).

[14] J. Carty, "Additional Force Required-Engineer", 8 Apr. 1909 (BLC).

[15] For a complete discussion of this phase of solid-state physics, see L. Hoddeson and G. Baym, "The Development of the Quantum Mechanical Electron Theory of Metals: 1900–1928", Proc. Roy. Soc. London **A371** (1980) 8–23 ; and L. Hoddeson, G. Baym and M. Eckert, "The Development of the Quantum-mechanical theory of metals, 1928–1933", Reviews of Modern Physics **59/1** (1987) 287–327.

[16] C. Weiner, "How the Transistor Emerged", IEEE, Spectrum **10** (1973) 24–33.

[17] "Electronic conduction in solids" and "Physics of the solid state", Case Reports, 1936-40 (BLC).

[18] G. Southworth, *Forty Years of Radio Research*, Gordon and Breach, New York, 1962, pp. 158–160.

[19] J. Scaff, "The Role of Metallurgy in the Technology of Electronic Materials", Metallurgical Transactions **I** (1970) 561–573; interviews by Hoddeson with R. Ohl and J. Scaff; also Ohl to Hoddeson, 2 Jan 1979.

[20] Transcript of filmed interview with Shockley, 1972 (BLC); W. Brattain, "Genesis of the Transistor", The Physics Teacher **6** (1968) 108–114; Brattain, "One Researcher's Personal

Account", Adventures in Experimental Physics **5** (1976) 3–13; W. Shockley, "The Invention of the Transistor: An Example of Creative-Failure Methodology", in: *Proceedings of the 2nd European Solid State Device Research Conference*, Lancaster, 1972, pp. 55–75 and Shockley "The Path to the Conception of the Junction Transistor", IEEE, Transactions **ED-23** (1976) 597–620; J. Bardeen, "Semiconductor Research Leading to the Point Contact Transistor", in: *The Nobel Lectures, 1942–1962*, Elsevier, Amsterdam, 1964, pp. 313–386.

[21] Interview by Hoddeson with D. Wooldridge.

[22] Ohl interview; Shockley, "Path" [20], 604.

[23] The following history is based on Bell Laboratories Notebooks of Shockley, NB 20455, Pearson, NB 20912; Bardeen, NB 20780; Brattain, NB 18194 and 21780; and Dreyer, NB 21373.

[24] Brattain to Hoddeson. 10 Jan. 1979.